Bringing Ecologists and Economists Together

Tore Söderqvist • Anna Sundbaum
Carl Folke • Karl-Göran Mäler
Editors

Bringing Ecologists and Economists Together

The Askö Meetings and Papers

Springer

Editors
Tore Söderqvist
Enveco Environmental Economics
Consultancy Ltd.
Oxholmsgränd 3
SE-127 48 Skärholmen
Sweden
tore@enveco.se

Anna Sundbaum
The Beijer Institute of Ecological
Economics
The Royal Swedish Academy
of Sciences
Box 50005
SE-104 05 Stockholm
Sweden
anna.sundbaum@gmail.com

Carl Folke
The Beijer Institute of Ecological Economics
The Royal Swedish Academy of Sciences
Box 50005
SE-104 05 Stockholm
Sweden
carl.folke@beijer.kva.se

Karl-Göran Mäler
The Beijer Institute of Ecological Economics
The Royal Swedish Academy of Sciences
Box 50005
SE-104 05 Stockholm
Sweden
karl@beijer.kva.se

ISBN 978-90-481-9475-9 e-ISBN 978-90-481-9476-6
DOI 10.1007/978-90-481-9476-6
Springer Dordrecht Heidelberg London New York

Printed on acid-free paper

Springer is part of Springer Science+Business Media (www.springer.com)

This book is dedicated to the memory of AnnMari and Bengt-Owe Jansson.

Foreword

As I have remarked before, it is no accident that the disciplines of *ecology* and *economics* have such similar names. Indeed, the names could easily have been interchanged. Both derive from the Greek, *oike*, which, as I am told, means, "household," or, "estate." The first known use of one of these terms is the still extant book, *Oeconomicus*, describing the management of a farm and the associated household (first half, fourth century B.C.). It was written by the Greek soldier and scholar, Xenophon. The term, "ecology", was not coined until 1866 (by Ernst Haeckel). Both terms refer to the analyses of complicated interacting systems in which individuals interact with each other and with the environment.

The Askö meetings indeed sought to bring these two disciplines together. But the motivation was not the search for intellectual parallels and mutual opportunities for learning. Rather, questions of policy and value dominated. The ecologists could perceive the dramatic negative consequences of the growth of the market economy – the effects on biodiversity inherent in the sheer spread of habitation and cultivation and in the monoculture found efficient in agriculture, the local and more widely diffused pollution of air and water with effects on humans and animals, and, at the most global scale, the emission of "greenhouse" gases from combustion. These are effects that ramify not only in space but over periods of time, centuries and millennia in the cases of climate change and nuclear waste disposal.

Strictly speaking, ecology is a descriptive study, telling us how living creatures and their environment interact and, in particular, what are the effects of human expansion on them. But it necessarily suggests values of conservation, and these values in turn lead to alternative policy proposals. These proposals inevitably have their costs. It may be proposed to exclude humans from living in otherwise desirable areas where game or vegetable biota would be destroyed, limit fishing and thereby the supply of fish for human consumption and sport, limit the use of energy or require restrictions, such as scrubbers, which divert resources which could otherwise be devoted to consumption and investment for the future. It is, of course, the analysis of alternative uses of scarce resources which has been the chief locus of modern analysis within economics.

The understanding of environmental issues within the main paradigms of modern economics is neither new nor peripheral; one may cite the magisterial, if at times soporific, studies of Arthur C. Pigou (first published in 1920), a professor at the University of Cambridge (and so at the then very center of economic analysis). Indeed, one of his chief examples was the smoke nuisance; the deficiencies of the unrestricted market system and proposals for remedying them were set forth clearly. Since then, the increasingly strong pressure of human economic growth on the biosphere and its documentation by ecologists has transformed the discussions from marginality to centrality within the world of economics.

The Beijer International Institute of Ecological Economics was created to bring the two scholarly traditions together, to inform the economists about the particular market failures and needs for collective action identified by the ecologists and to bring to the ecologists the particular views of economists as to the costs of actions and as to the choices of actions which will minimize costs. Karl-Göran Mäler, with the collaboration of Carl Folke, Partha Dasgupta and others deserve the credit for bringing this concept to fruitful and effective being.

In particular, they had the idea of bringing ecologists and economists together, not only in specific collaborations but also in informal sessions. Indeed, the broad vision of AnnMari and Bengt-Owe Jansson gave a setting to the meetings, the marine research station at Askö, in the Stockholm Archipelago. The cycles of nature dominate the setting, a very small island surrounded by other islands, a research station with equipment and charts denoting the changes in Baltic fisheries, meetings interrupted as the more knowledgeable attendees identify various birds, yet resuming on some technical point on which the economists are trying to understand their ecological colleagues or vice versa.

What became clear very quickly is that there was not merely a common interest in formulating values and policies. There was also a sense of the common elements in the two disciplines. Terms like "portfolio" or "discounting" began to appear in the papers of the ecologists, while economists grappled with "resilience" and learned that price systems might fail to avert catastrophes in fairly straightforward models of pollution of a lake by phosphates.

The Askö meetings have become an institution of central importance in this interdisciplinary development. The participants, with considerable continuity and yet steady input of new members, have learned much from each other. The statements issued and included in this volume represent widely-understood distillations of accepted knowledge. The process of arriving at these consensuses has been of equal significance.

Kenneth Arrow

Preface

Askö: Building Cognitive and Non-cognitive Capital in the Public Interest

Over 40 years ago when I entered graduate school I was told, along with the other new students, about the academic life. The message was roughly the following:

> Faculty are barnacles on the ship of productivity. They are privileged to enjoy the luxuries of the academic life.

A few years later John Krutilla put things in a different perspective. Academics who work for tax exempt institutions, in his view, had a social responsibility to contribute to public well being. Over the intervening 4 decades I have encountered more academics who feel like John than those who view their positions as a privilege without social responsibilities. Most of us try to contribute and hopefully our results do, in muddy and meandering ways. This volume presents the results of a group of academics who *really did make a difference*!

There is widespread recognition that much important research starts at the boundaries of the existing disciplines. University administrators and research foundation staff have coined a number of terms to describe these activities. Interdisciplinary, multidisciplinary and transdisciplinary are three that come to mind (I won't attempt to explain perceived distinctions between them). These academic leaders increasingly want authentic efforts on important boundary research questions to happen but don't know how to get started. Nowhere is this need more apparent than in the global environmental challenges we face today.

Karl-Göran Mäler discovered the secret for ecologists and economists. He recognized that real collaboration requires not simply the physical proximity of exceptionally talented scientists from these disciplines. Their cognitive skills are important but not enough. Genuine collaboration needs a culture that encourages these academics to be adventurous and think outside the domain of their expertise. It must be tolerant and constructive. In such a setting, respect and admiration build trust and friendship. He established a process that built a culture for the development of non-cognitive skills in groups. These skills relate directly to research at the boundaries of these disciplines. They reminded me of the recent research on why learning and ability depend in complex ways on an individual's genetic endowment and environment.

More specifically, Heckman (2007) has argued that the formation of cognitive and non-cognitive skills is important to the development of human capital and ability. He describes a dynamic model for children's development and extensions to the early human capital literature on learning and ability that distinguishes "self productivity" and "dynamic complementarity". The former describes a process for how an individual's capabilities produced at one stage enhance the capabilities attained in later stages. His concept of dynamic complementarity refers to the synergistic role of capabilities achieved at each stage of an individual's life in promoting future accomplishments.

It seems to me that Heckman's model is an ideal metaphor for what the group of ecologists and economists has achieved at Askö. In the years since the meetings began in 1993, they have demonstrated it is possible to transcend disciplinary limitations and exploit the joint insights of both disciplines. These meetings and the set of researchers that shared those experiences provide an example of how groups can create non-cognitive capital that allows them to learn outside their disciplines. The results reported in this volume clearly confirm that judgment.

What lessons can we draw from the experiences for new initiatives that seek to create group cultures that support and sustain research bridging disciplines sharing common boundaries and facing the need to work together for genuine progress? Here are few:

- Select senior scholars with outstanding credentials who are intellectually adventurous.
- Commit to a sustained sequence of meetings and be patient for results.
- Assign responsibility and authority to leaders in each discipline to launch the annual discussions.
- Use annual outputs to track progress.
- Isolate the group while they meet.
- Keep logistics of meetings seamless for all who participate.
- Clone Karl-Göran Mäler!

V. Kerry Smith

Reference

Heckman, J. J. (2007). The economics, technology, and neuroscience of human capability formation. *Proceedings of the National Academy of Sciences, 104*, 13250–13255.

Acknowledgments

We thank the Anna-Greta and Holger Crafoord Foundation for financial support for this book.

Contents

Contributors

Kenneth Arrow
Department of Economics, Stanford University, USA
arrow@leland.stanford.edu

Edward B. Barbier
Department of Economics and Finance, University of Wyoming, USA
ebarbier@uwyo.edu

Scott Barrett
School of International and Public Affairs & Earth Institute,
Columbia University, USA
sb3116@columbia.edu

Stephen Carpenter
Center for Limnology, University of Wisconsin, USA
srcarpen@facstaff.wisc.edu

Anne-Sophie Crépin
Beijer Institute of Ecological Economics, the Royal Swedish Academy of Sciences, Sweden
annesophie.crepin@beijer.kva.se

Partha Dasgupta
Faculty of Economics, University of Cambridge, UK
partha.dasgupta@econ.cam.ac.uk

Paul R. Ehrlich
Department of Biological Sciences, Center for Conservation Biology,
Stanford University, USA
pre@leland.stanford.edu

Carl Folke
Beijer Institute of Ecological Economics, the Royal Swedish Academy of Sciences, and Stockholm Resilience Centre, Stockholm University, Sweden
carl.folke@beijer.kva.se

Lance Gunderson
Department of Environmental Studies, Emory University, USA
lgunder@emory.edu

Claude Henry
School of International and Public Affairs, Columbia University, USA and Laboratoire d'Econométrie, Ecole Polytechnique, France
ch2396@columbia.edu

Terry Hughes
ARC Centre of Excellence for Coral Reef Studies, James Cook University, Australia
terry.hughes@jcu.edu.au

AnnMari Jansson
Department of Systems Ecology, Stockholm University, Sweden

Bengt-Owe Jansson
Department of Systems Ecology, Stockholm University, Sweden

Ann P. Kinzig
School of Life Sciences, Arizona State University, USA
kinzig@asu.edu

Simon Levin
Department of Ecology and Evolutionary Biology, Princeton University, USA
slevin@princeton.edu

Karl-Göran Mäler
The Beijer Institute of Ecological Economics, the Royal Swedish Academy of Sciences, Sweden
karl@beijer.kva.se

Shahid Naeem
Department of Ecology, Evolution and Environmental Biology, Columbia University, USA
sn2121@columbia.edu

Elinor Ostrom
Workshop in Political Theory and Policy Analysis, Indiana University, USA
stodd@indiana.edu

Charles Perrings
School of Life Sciences, Arizona State University, USA
charles.perrings@asu.edu

Stephen Polasky
Department of Ecology, Evolution, and Behavior, University of Minnesota, USA
polasky@umn.edu

Alison G. Power
Department of Ecology and Evolutionary Biology, Cornell University, USA
agp4@cornell.edu

Robert J. Scholes
Council for Scientific and Industrial Research, South Africa
bscholes@csir.co.za

V. Kerry Smith
Department of Economics, W.P. Carey School of Business,
Arizona State University, USA
kerry.smith@asu.edu

David A. Starrett
Department of Economics, Stanford University, USA
davstarret@aol.com

Thomas Sterner
Department of Economics, University of Gothenburg, Sweden
thomas.sterner@economics.gu.se

Anna Sundbaum
The Beijer Institute of Ecological Economics, the Royal Swedish Academy of Sciences, Sweden
anna.sundbaum@gmail.com

Tore Söderqvist
Enveco Environmental Economics Consultancy Ltd, Sweden
tore@enveco.se

Jeffrey R. Vincent
Nicholas School of the Environment and Sanford School of Public Policy,
Duke University, USA
jeff.vincent@duke.edu

Brian Walker
CSIRO Sustainable Ecosystems, Canberra, Australia
brian.walker@csiro.au

Part I
Introduction

Chapter 1
The Askö Experience

Tore Söderqvist and Anna Sundbaum

A multitude of meetings with other scholars are usually found in a scientist's calendar, and some scientists even have many meetings with people representing other scientific disciplines. How can such interdisciplinary meetings be made productive and rewarding to the participants? This is a challenge, because differences in scientific traditions, ways of thinking and jargons, and tendencies to strong territoriality, might make it hard for the participants to understand each other. One strategy for trying to solve this problem might be to prepare such a meeting carefully. This might include the development of a detailed agenda that makes sure that there is a fair distribution of time and chairmanships among the different disciplines, so that every participant feels comfortable in advance with the degree of attention that she/he and her/his discipline will get. However, this book is about a series of meetings that are characterized by a quite different strategy: The Askö meetings.

These meetings are named after the place where they are usually held, i.e., the island of Askö about 70 km south-west of Stockholm, Sweden. The first Askö meeting was held in 1993 on the initiative of the Beijer Institute, one of the research institutes of the Royal Swedish Academy of Sciences. At that time, the institute had quite recently been reorganized as an institute of ecological economics. By definition, ecological economics is about co-operation among scholars from different disciplines and a crucial task for the institute was to stimulate such co-operation. The Askö meetings turned out to be one successful means for this. Since 1993, there have been one or two Askö meetings annually. Not all of them has taken place on the island of Askö, but they are usually referred to as

T. Söderqvist
Enveco Environmental Economics Consultancy Ltd,
Oxholmsgränd 3, SE-127 48,
Skärholmen, Sweden
e-mail: tore@enveco.se

A. Sundbaum
The Beijer Institute of Ecological Economics, the Royal Swedish Academy of Sciences,
Box 50005, SE-104 05, Stockholm, Sweden
e-mail: anna.sundbaum@gmail.com

T. Söderqvist et al. (eds.), *Bringing Ecologists and Economists Together*,
DOI 10.1007/978-90-481-9476-6_1, © Springer Science+Business Media B.V. 2011

"Askö meetings" as well, since after a few years "Askö" did not only refer to a geographical place but also indicated a particular way of having a meeting.[1] Most Askö meetings have resulted in a publication, and these are referred to as Askö papers.

At the Askö meeting in 2005, we felt that the time had come to look back at what the meetings had accomplished so far and also have a kind of evaluation – what (if anything) makes the Askö meetings special? The result is this book, which has the purpose to put both the Askö meetings in themselves and their results in terms of publications in perspective. The former is done by Kerry Smith's preface and Kenneth Arrow's foreword and six chapters in which Karl-Göran Mäler, Partha Dasgupta, Paul Ehrlich, Simon Levin, AnnMari Jansson and Bengt-Owe Jansson, and Carl Folke give their views on the Askö meetings, and the latter is accomplished by reprints of the first ten Askö papers, each accompanied by two commentaries written by authors who represent different scientific disciplines – in most cases ecology and economics, respectively. The commentaries indicate that reading the Askö papers will be rewarding not only because they say something historically about what ecologists and economists once agreed upon to write, but also because the subjects brought up in the papers tend to be highly relevant also today. This is also true for the policy advice given in some Askö papers, which suggests that implementing policies according to this advice is even more urgent today.

The Askö meeting period covered by this book is the first 10 years, i.e., 1993–2002. Table 1.1 gives an overview of the topics of the 12 meetings held in this period. Ten of these meetings resulted in publications, which all are reprinted in this book. Nine of these papers were published in international scientific journals and one is available in the discussion paper series of the Beijer Institute. The table also shows that the number of participants in these meetings varied between 10 and 22 (average: 16). The total number of participants is 194, but this sum mostly consists of people who participated in at least two of the 12 meetings in Table 1.1. 35 of 194 refer to scholars who participated in only one Askö meeting in this period. For details, see the appendix, which contains lists of participants for each Askö meeting.

Several of the most frequent Askö meeting participants give their view on the meetings in chapters in this book. In order to complement their texts, the rest of this introduction contains lists of answers and citations[2] from some of the questions in a questionnaire that was e-mailed to all the 76 persons who have participated in at least one Askö meeting in the period of 1993–2005. Twenty-nine persons answered the questionnaire partly or wholly and 20 of them have participated in more than one meeting. Thirty-eight percent is a small response rate, so what follows should

[1] By the way, the letter "ö" is frequent in this book. While Swedes, Finns, Germans and a few other nationalities know how to pronounce an "ö", most other people do not. Hint: In French, "ö" is the same sound as "œ". In English, the vowel sound in the word "girl" is almost pronounced as an "ö".

[2] A few citations have been slightly edited for increasing anonymity.

Table 1.1 An overview of the 12 Askö meetings held in the first years (1993–2002)

Year	Topic	Number of participants	Publication
1993	(No specific topic)	15	Yes, in *Bioscience*
1994	Economic growth and the environment	11	Yes, in *Science*
1995	The resilience of social and biological systems	17	Yes, in *Environment and Development Economics*
1996	The environment and ethics	16	No
1997	Population, food availability and the environmental resource base	16	Yes, in *Science*
1998	Valuing ecosystem services	17	Yes, in *Science*
1999	Institutions for managing ecosystem services	10	Yes, in *Environmental Science and Technology*
1999	Evolution in biological, social and economic systems	19	No
2000	Consumption	12	Yes, in *Journal of Economic Perspectives*
2000	Environment, development and trade	20	Yes, in *Resource and Energy Economics*
2001	Indicators	19	Yes, as a Beijer Institute discussion paper
2002	Dealing with uncertainty: Policy making in the twenty-first century	22	Yes, in *Ambio*
Total		194	Ten publications

be interpreted as a collection of examples of opinions rather than answers that together are representative for the whole group of participants.

We prefer to let the answers in the lists below speak for themselves and urge the reader to go through them. Taken together with the chapters about the meetings, we hope that the reader will get an understanding for the special character of the Askö meetings and the pros and cons of the Askö strategy for creating interdisciplinary co-operation.

Q1. Are There Any Unique Features of the Askö Meetings in Comparison to Other Scientific Meetings You Have Participated in?

A. Yes: 28. No answer: 1

What features?

(1) No agenda! (2) No presentations. (3) Though I have been to many interdisciplinary meetings, this is the only one that seeks to bring these two particular disciplines together. (4) Meetings do not get friendlier than this.

The ambition of the meeting: Put together a bunch of well-qualified researchers in very different disciplines and make them brainstorm during 2 days about a given question with ambition to write a paper about the topic and publish it in a top journal. My experience on the potential for cooperation between different scientific disciplines was that there is a need for a rather long start up period where people learn each other's way of thinking. In the Askö meetings I attended, most people were already used to interacting between disciplines so people could start thinking about the problem from the first moment.

There is simply nothing like the Askö structure: Participants are notified of the general subject to be discussed, but are presented with no paper, nor a framework on which to work. There is no agenda, other than the requirement of having to sit and discuss.

Much more informal than most meetings; good interdisciplinary interaction; sufficient time to discuss each issue raised.

No agenda – just an opening statement or topic. High quality people who look to listen to alternatives. Beautiful place, good food, wonderful sauna!

Coming to a small research station in the middle of the Baltic archipelago the participants have cut the ties, entangling them in the trivia and conventional networks of civilization. The mind is open for new stimuli, new approaches to old problems, discussions free from professional prestige. Everybody is conditioned to the same mental level. The true nature shows its beauty and power.

The relatively free-flowing nature of the discussion.

No other meeting brings together, on a continuing basis, leaders from both communities. The continuity in the membership has built trust, and encouraged convergence in viewpoints. The format is superb and effective.

The unstructured format and the mix of ecologists and economists.

The fantastic isolated environment without other distractions.

The meetings stimulate a free and open discussion without any background documents, no presentations or PowerPoints. I was rather skeptical before the first meeting, but I am now totally convinced that the meetings provide a rare platform for natural and social scientists to discuss important topics. The track record of publications is certainly a testimony to the value of the exercise.

In depths (high quality) interdisciplinary conversations on focused problem with short paper as product.

Good interaction between ecologists and economists.

(continued)

Q1. (continued)

The unusually informal and creative atmosphere.

The relaxed atmosphere and the positive attitude among participants.

While there is a general theme, there is no agenda. A couple of people (usually an economist and an ecologist) are asked to make introductory remarks and after that the meeting evolves.

The uniformly high quality of the participants and the sustained high level of the discussions are unusual. Meetings draw the best scholars in environmental economics and ecology and put them in a setting that guarantees fruitful interchange.

The high caliber of the participants, the interdisciplinary nature of the meeting, the relaxed and informal way that people exchange opinions and ideas about a scientific topic, and that way that at the end there is convergence to a few main points and a draft for a paper.

The Askö meeting gathered fantastic people who have a deep desire to work together to contribute towards solutions of hard social problems involving economics and ecology.

The meeting involved spontaneous discussion of a topic with no major preparation in advance. This increased spontaneity, but made the goal of producing a publishable product more challenging, because the discussion had a stream-of-consciousness trajectory.

Bringing together top-scientists from various fields to address and discuss key questions for the fate of the Planet is of course extremely giving and may produce major impact and influence to policy makers and others. It is discussed in an informal atmosphere, which brings forward ideas that otherwise might be hard-fetched.

Informal yet high-quality meeting between top-level scientists from different disciplines.

My additional experiences are from meetings mainly at the NCEAS [National Center for Ecological Analysis and Synthesis] and ICSU [International Council for Science]. The difference I noted was that at the Askö meeting I took part in it was already decided who would be the lead authors. This saved a lot of time because lead authors were well prepared and we did not need any discussion about who would write.

Multidisciplinary approach.

It was unusually informal. I have been to other meetings with as few participants and in as isolated a location, but I have not been to any that both did not require any preparation and did not involve any formal presentations.

The meeting was small scale and highly focused on a single topic. That allowed each of us to contribute to that common topic from our own perspective.

The unstructured format, and the mix of world-class ecologists and economists.

(continued)

Q2. Did the Meetings Change Your View on the Potential for Cooperation Between Different Scientific Disciplines (Such as Ecology and Economics)?

A. Yes: 15. No: 11. Yes and no: 2. No answer: 1

If yes, how?	If no, why?
In the early days, the Beijer Institute was truly revolutionary in trying to bring disciplines together. What I learned was that this was possible, but that it required a sustained effort and patience.	I came into this expecting that true interdisciplinary research required each person spending a lot of time learning the other discipline. I still believe this is true.
My experience on the potential for cooperation between different scientific disciplines was that there is a need for a rather long start up period where people learn each other's way of thinking. In the Askö meetings I attended, most people were already used to interacting between disciplines so people could start thinking about the problem from the first moment.	I already had a fairly positive view; the meetings confirmed it.
With enough time spent to discuss, economists and ecologist can learn a lot from each other.	I have always been an enthusiast about inter-disciplinary work – the meetings help to confirm my views but not change them.
More similarities between the two than I expected. But the differences need to be the focus: e.g., multi stable states, at least three speeds of variables, surprises in ecosystems vs. focus on equilibrium, on single or two variables predictable. Those differences started to be addressed, and effectively.	I knew it could be done – but it was nice to participate in a few living examples!
	Already a firm believer.
	It rather confirmed my belief that this potential is great.
	Already know about this potential from my position at Beijer.
	I have engaged in multidisciplinary conversations and collaborations for much of my career. The current climate between economics and ecology is warm in comparison with 2 decades ago, and I am optimistic about the promise for dialogue and mutual understanding.
	My views had already been shaped by working with Beijer Institute people (i.e., Askö folks), so that's why I answered both "yes" and "no".
	I'm a firm believer that people from different fields can cooperate, being a bit of a naive optimist.

(continued)

Q2. (continued)

If yes, how?	If no, why?
The lack of personal prestige in the informal atmosphere and the big portion of humor, coloring the discussions, softened the defense of personal opinions and favorite theories. I give the open atmosphere and the many laughs much credit for the successful results. Ecology and economy do go together. It is to great extent a matter of willingness to listen and to understand.	I am already convinced that there is great potential (and need) for collaboration between different scientific disciplines. In practice though, it is not so easy, partly because of the funding system, at least in Sweden.
By demonstration, and spin-off activities.	I have long been a passionate believer in the value of interdisciplinary work. It reinforced my belief, but it didn't change it.
Although I have followed the collaboration between economists and ecologists at a distance, it was only when I joined the Beijer Board that I got involved personally. However, although we have had a few non-economists/ecologists, it has not tested natural and social science interaction in a broader context. The use of models in ecology and economy has provided a common language for the discussion. This would not be as easy when you move outside these disciplines.	I had participated in several previous Beijer Institute events, and I was already convinced of the potential for cooperation between ecologists and economists as a result of that participation.
Initial deep differences between disciplines are lessened by a developing understanding of where the other discipline is coming from. But it takes time. The continuity of a core of participants at Askö meetings has allowed this to occur.	
I became aware of the links and the common objectives between ecology and economy.	

(continued)

Q2. (continued)

If yes, how?	If no, why?
I really think that activities like Askö have helped enormously at getting economists and ecologists to work together. It was my first meeting of a group of economists and ecologists that had extensive experience working together. They had proceeded beyond the initial stages of sorting out how to work together and communicated very effectively. The meeting allowed me to better appreciate the similarities and differences between questions and approaches in ecology and in economics, and to see concrete issues and ways to address them at the interface between ecology and economics. We need more interaction. It made clear to me that many of the general results in game theory had been reached by economists well before they were appreciated by biologists. They proved not only that ecologists and economists could work together, but that, by working together they could take their own disciplines forward.	

Q3. Did the Meetings Change Your Own Research in Any Way?

A. Yes: 17. No: 10. Yes and no: 1. No answer: 1

If yes, how?	If no, why?
I was always inclined towards interdisciplinary research, but the Askö meetings made me even more confident that this was the right course.	The topics of my research are probably not very ecologically oriented.
But for the Askö meetings I doubt whether during the past 12 years or so I would have devoted so much of my research to helping to shape ecological economics. Interactions with ecologists meant that I got to know them personally, which in turn meant that I collaborated with some of them on non-Askö projects.	I've only been to two [meetings], and knew many of the participants already. Thus, while I found them intellectually stimulating, they did not change my research in any fundamental way.
More attention to the significance of human expectations and slowness (compared to ecosystems) of response to increasing rigidity and to recovery from traps.	Unfortunately I am no longer an active scientist doing research, but I have benefited greatly in my work as a science administrator.
To some extent. We got more respect for economics as a stringent science and the tremendous strength of the market economy. We learned also to formulate our opinions on the use of natural resources in a way more adequate for economic scrutiny.	I was invited to discuss what I was already doing. But, I learned a lot and took home with me.
I have learned much, and now incorporate economic components often in my papers.	Already collaborate with economists.
	Essentially the same answer as to Q2 [My views had already been shaped by working with Beijer Institute people (i.e., Askö folks), so that's why I answered both "yes" and "no"]. I have been doing joint research with some of the Askö "regulars" for some time.
	It did not create major opportunities for cooperative work but strengthened my conviction that this is an area worth exploring.
	I learned a few new things but I would not say it changed my research in any specific way.
	My research was already multi-disciplinary.
	Not yet, anyway. The meeting did not get into much depth on the main issues I research. It did, however, prompt me to think more about the political economy of policy responses to environmental problems.

(continued)

Q3. (continued)

If yes, how?	If no, why?
Carried out research on hyperbolic discounting. I became more familiar with the concepts of resilience and have been writing quite a bit about resilience and robustness of social-ecological systems since then. Hard to say – subtle changes due to new ways of thinking. I learnt new things to consider in my research and was strengthened in my belief that interdisciplinary research is worthwhile. They have increased awareness of the need for research on valuing both non-market and marketed goods and services in ways that reflect ecological reality. The meetings I have attended have been wide ranging and stimulating. They have also helped me make connections with new people with whom I have sympathetic research agendas. I extended my research interests to sustainability, economic/ecological modeling and spatial modeling. Essentially the same answer as Q2 [I really think that activities like Askö have helped enormously at getting economists and ecologists to work together]. I have been doing joint research with some of the Askö "regulars" for some time.	

(continued)

Q3. (continued)

If yes, how?	If no, why?
It made me realize that incorporation of economic perspectives is a lot more straight-forward than I realized, so I have felt less hesitant to make pursue research at this interface.	
It inspired me tremendously.	
It interested me in evolutionary and behavioral economics, and several years later I read the work of Bowles, Gintis, Fehr, and Gigerenzer as part of writing a book.	
I became much more aware of the notions of multiple equilibria, tipping points and what is now called punctuated equilibria as well as the use of adaptive management as a desired strategy rather than a default strategy.	

(continued)

Q4. In General, What Were the Most Important Pros and Cons with the Meeting(s)?

Pros Quality and mix of the people attending; the friendly atmosphere.
Cons I would be inclined to give the meetings more structure, and to focus more on controversies than topics. This would give more direction to our conversations, exploit the talents around the room, and make it easier to publish the paper resulting from the meeting.

The most important *pros* are the opportunity to interact with people from other disciplines and try to find solutions to important problems that cannot be solved within one discipline. One of the most important *cons* is the lack of women and younger people among the participants.

There ARE no cons. The Askö meetings are wholly original and productive.

Pros See answer to Q1 [Much more informal than most meetings; good interdisciplinary interaction; sufficient time to discuss each issue raised].
Cons The unorganized structures of the meetings sometimes lead to inefficient use of time.

Pros The results in terms of publications.
Cons Some of the participants tend to dominate and do not listen enough to other participants. This hampers creativity.

Pros Open agenda, good balanced people, fun on islands!
Cons One meeting when none of those positive features were evident.

Pros To get personal contacts with leading economists for further collaboration. To get some insight of how economists are thinking. To get opportunity to explain the complexity of nature, the importance of feedback loops, ecosystem services, biodiversity and the instability of ecosystems.

Most important *pros* are the energy and intelligence of the participants, the free-form nature of discussion, and the chance for new thinking. Perhaps the biggest *con* is that it largely involves old friends and like-minded thinkers. A great experience for them, but participation could be a bit wider.

Pros are the choice of people, and the continuity of membership. No real *cons*, except perhaps longer meetings would be good occasionally.

The most important *pro* is the broad perspective that derives from having such a diversity of people. The most important *con* was the occasional lack of focus.

(continued)

Q4. (continued)

Outstanding scholars invited, good long times for group discussions as well as opportunities for smaller groups to get together and argue through important points and possible disagreement. I do not have any(!) negative reactions at all – except that I would like to come back at some time in the future!

The *pros* are listed under Q1 [The meetings stimulate a free and open discussion without any background documents, no presentations or PowerPoints. I was rather skeptical before the first meeting, but I am now totally convinced that the meetings provide a rare platform for natural and social scientists to discuss important topics. The track record of publications is certainly a testimony to the value of the exercise.]. I do not have many (any?) *cons*, except possibly that the format is especially suited to a relatively small group of people that have previously collaborated. However, we have also brought in "outsiders" and it worked. Some of the publications have taken a very long time and it is a bit haphazard if you get the right intellectual environment that will stimulate a few people to take the lead in preparing the manuscripts. Also, some of us have been more or less free-riders on the publications.

Pros Super people, enough time for in-depth talks, useful product.
Cons A long trip.

High level people who are very committed. Free atmosphere.

Pros A unique opportunity to meet and interact with top-level ecologists and economists. The informal interaction, no arrogance (sometimes the latter might have existed but was well hidden).
Cons Too static composition of participants. This makes it harder for "newcomers" to participate on equal terms. Too little efforts to involve new generations of ecologists and economists.

Pros The loose agenda that leave room for creative thinking.
Cons Maybe the persons being rapporteurs should get some more time before the meeting for preparation.

Pros Stimulating and informative; great fun; development of a much better understanding of economic strengths and weaknesses; publication of several important papers in key journals that have had a significant impact on ecological economics.
Cons (i) While acknowledging the value of continuity in participants, there could have been a better mix of new people – more of them in each meeting. Rotating some of the core would be good. I think some of the best meetings were in the early years, and with time the core group has

(continued)

Q4. (continued)

delivered what it knows and loves best, and there's a need for a much bigger proportion of new people, with different expertise and experience. (ii) It might be more beneficial if participants were better prepared and informed on the topic before the meetings.

I have enjoyed both Askö meetings I attended, without reservation. The settings have been beautiful, the accommodations, food and libations generous, and the meetings and interaction outside of meetings very stimulating. I think that the whole Askö style, of an informal setting, with plenty of time to talk, and with the right mix of people, is very successful. And I can't praise enough the high caliber of the participants and of the discussion – a perfect mix of disciplinary diversity, research interests, and personality types. I always come home from these meetings full of energy and stimulation and thinking about new questions and approaches.

Pros The quality of the participants, their willingness to cooperate and the informal atmosphere allowing everybody to express his/her opinion.

Cons Sometime it was difficult to convergence if the topic was not very well defined. The less senior participants do not talk as much as they should.

It's hard to find anything to complain about. Great people – great science. Who could ask for more?

The greatest benefit came from the opportunity for open discussion with an incredible group of people who had great ideas and a history of conducting discussions like this. They could tackle a new topic with which they had little familiarity and, with the help of a couple subject matter experts, make important contributions. The biggest challenge (not necessarily a disadvantage) was the lack of structure or clear conclusions to the discussion – yet with the expectation that this would lead to a well-organized insightful publication.

The most important *pro* for me as an ecologist was the opportunity to meet and interact with top-level scientists from economics. The most important *con* was the lack of clearly identified products and follow-up.

I like the concept of bringing together top scientists to discuss common themes. This is almost impossible to do from your offices (by e.g. phoning or e-mailing).

Pros Interactions with economists.

Pros Pleasant small group interactions in a lovely and isolated location.

Cons Dominant Americans, lack of concrete agenda.

(continued)

Q4. (continued)

Pros	Scholars from different disciplines working *together* productively on a single product.
Cons	My attendance at that meeting was a one-shot occurrence so that experience was not reinforced for me.
Pros	Assembling a group of smart people in a relaxed environment that encouraged brainstorming. Fantastic work by the rapporteurs, in assembling a rough draft of the manuscript by the end of the meeting.
Cons	In view of the topic we came up with, a couple of political scientists would have enriched the discussions considerably.

Part II
The Askö Meetings

Chapter 2
A Brief History

Karl-Göran Mäler

Go back 20 years and try to remember the attitudes of economists towards ecologists and vice versa. In general, these two groups were quite hostile to each other, ecologists accusing economists for being responsible for the increasing human encroachment into the natural environment (pollution, destruction of habitats, etc.) and economists replying that ecologists should stay where they belong, studying nature and not trying to build a "new" economics. Both sides in the conflict did have good arguments, but the "war" could hardly be judged as productive. Bengt-Owe Jansson, professor of Systems Ecology at Stockholm University, and I discussed the situation in the end of the 1980s and we were both convinced that we had to find new ways of getting ecologists and economists together. There had, of course, been previous attempts to try to get economists and ecologists together, for example, a workshop in Saltsjöbaden (in the Stockholm archipelago) 1982 on *Integration of Economy and Ecology – an Outlook for the Eighties*. However, nothing was really accomplished at that meeting; only more demonstrations that ecologists and economists could not understand each other.

In the early 1990s a new opportunity was created as the Royal Swedish Academy of Sciences invited its members for suggestions for the scientific content in the "new" Beijer Institute.[1] Bengt-Owe and I proposed an institute focusing on cooperation between ecologists and economists and eventually the Academy decided along with Bengt-Owe's and my proposal. A year later, the Academy appointed me as director of the new Beijer Institute and I invited Carl Folke to join me as deputy director.

[1]The Beijer Institute was created in 1977 as an institute for carrying out research in energy and human ecology and funded by the Kjell and Märta Beijer Foundation. In 1989, the Royal Swedish Academy of Sciences and the Swedish Government agreed to transfer the Institute into a new Government agency, the Stockholm Environment Institute. However, the Beijer Foundation was willing to continue its support to the original Beijer Institute, and the Academy had to decide on its direction.

K.-G. Mäler
The Beijer Institute of Ecological Economics, the Royal Swedish Academy of Sciences,
Box 50005, SE-104 05, Stockholm, Sweden
e-mail: karl@beijer.kva.se

T. Söderqvist et al. (eds.), *Bringing Ecologists and Economists Together*,
DOI 10.1007/978-90-481-9476-6_2, © Springer Science+Business Media B.V. 2011

Carl (most often called Calle) had that year finished his PhD and was spending some time in Boston when I faxed him asking him to join me and he almost immediately accepted the position.

Calle and I discussed various ways of getting ecologists and economists together in a productive way and we came up with a number of ideas. One, which I decided upon very early, was that we should not have general discussions between economists and ecologists. From what I learned from the symposium in Saltsjöbaden, I was convinced that those discussions would be producing even further bad feelings between the two professions. Instead I was interested in having concrete research projects in which economists and ecologists would have their natural jobs. Thus we started the Baltic Drainage Basin Project (funded by the Swedish Environmental Protection Agency and the European Union), which included marine ecologists and economists from Germany, Poland, United Kingdom and Sweden. We were also looking further to other research programs (the next was the Biodiversity Program, directed in an excellent way by Charles Perrings, and a third program was a program on Property Rights and the Environment, directed by Susan Hanna). Then, one evening when Calle and I were discussing other possible activities, we came up with the idea to organize special meetings at Askö Marine Laboratory. This lab belonged (and still belongs) to the Stockholm University and is located on an island in Trosa archipelago south-west of Stockholm. It is used by the Department of Systems Ecology for undergraduate training and as a research lab for PhD students and more senior researchers. Being located in the archipelago, bird watching could be rewarding and in particular the White-tailed sea eagle could be seen on the island. Director of the lab at that time was Bengt-Owe and we knew that he would give us permission to get access to the lab for Beijer activities.

In order to be successful, I decided (together with Calle) on three points:

1. We had to invite a selected group of people, who we could expect the share our view of the necessity for economists and ecologists to cooperate.
2. The meeting should be very informal – in fact so informal that we had no chairman and the participants had to catch the floor him or herself. I decided on the theme for the discussion and I also asked two participants to act as rapporteurs (if my memory is correct, Partha Dasgupta and Calle volunteered to be rapporteurs at the first meeting) and I also gave a very brief background to the theme and then the discussion started.
3. I was also very keen on creating a warm atmosphere so that the participants would be attached to the idea of being together people with different scientific backgrounds. Thus, we spent the first evening eating, wine drinking and singing and for some of us in the sauna.

Obviously, we also wanted a product out of the meeting such as a preliminary manuscript for later publication. This manuscript was then circulated among the participants for comments, changes, and general improvements. Eventually, we had a manuscript in such a form that I believed we could publish it.

There has been a core group of persons who have been invited regularly and also participated regularly – Kenneth Arrow, Partha Dasgupta, Paul Ehrlich, Simon Levin and Brian Walker.[2] In the beginning, not all members of the Beijer Board were invited (mainly due to limitation of space and the need for some external inputs to our discussions) but that has changed over time. Later, all Board members were invited (there are not as many Board members now as there were in the initial Board).

One important reason for productive Askö meetings was the social components. During walks around the island, groups of participants were able to know each other better and to understand the others' points of views. In the evening, we all had good dinner (always the same – Swedish hash, smoked salmon with boiled potatoes, cheeses and coffee) with ample supply of wine. After singing, the evening ended (at least for some brave persons) with a visit to the sauna and a compulsory dive into ice cold Baltic.

Looking back of the 12 Askö meetings this book is about, we were successful in most of the cases. Only in two cases we were not able to produce a manuscript (that is, we could not agree on the preliminary manuscript). In most cases we were also able to publish the manuscripts: three in *Science* and six in other journals, like for example *Journal of Economics Perspectives*, *BioScience* and *Environment and Development Economics*.

We felt very proud of what we accomplished at Askö so we decided to create new meetings, building on the same idea. Thus we organized one meeting at Resources for the Future in Washington DC, which resulted in a paper on economic-ecological modeling and another one at Stanford University, which resulted in the article "Are we consuming too much?" published in *Journal of Economic Perspectives*. However, it turned out that the participants in the Askö meetings preferred the more primitive environment at the Askö Laboratory and it became too expensive to organize the meetings in the United States, so our excursions to other parts on earth for our "Askö" meetings came to an end.

A major reason for the success of the meetings were the role played by the staff, who made the preparations for the meeting, making all the logistics function effectively, cook our meals, and force us to follow the timetable. Christina Leijonhufvud has been involved since the start. She was for the first 7 years accompanied by Astrid Auraldsson. When Astrid left Beijer, Anna Sundbaum replaced her. They have all served in a rather difficult operation but they have solved all problems that occurred[3]

[2]Kenneth Arrow, a brilliant economist and Nobel Laureate, and Paul Ehrlich, a brilliant ecologist and receiver of as many prizes as there are in ecology, of course knew each other, being professors at the same university, Stanford University, but they had never discussed scientific issues with each other. Through the Askö meetings, they became good friends and were able to produce high quality work. Brian Walker and Simon Levin have similarly cooperated with economists and produced many articles written jointly with economists (with, for example, the current author and with Kenneth Arrow, Partha Dasgupta, Tasos Xepapadeas and Aart de Zeeuw).

[3]For example, once going back from Askö, the bus that was expected to take us back to Stockholm, was not there. Some of the participants had flights from Stockholm in the afternoon, and they were beginning to be worried about missing their flight. Christina and Anna were able, not only to get sufficient number of taxis in very short time but also forcing the bus company to accept the liability for the taxis.

very efficiently. The Beijer Institute is very grateful for their services. Thank you Christina, Astrid and Anna.

We have had many scholars participating in the meetings. Let me single out three, who are not any longer with us – AnnMari Jansson, Bengt-Owe Jansson, and Bert Bolin. All three took an active part in our discussions. Bengt-Owe and AnnMari were our host and hostess at Askö and they knew anything worth knowing about the island's ecology, birdlife and history. Bert, as always an intellectual giant, played an important role for pulling our, sometimes flippant, discussions back to order. They will be sorely missed.

Chapter 3
Why and How?

Partha Dasgupta

Intellectual life moves at a fast pace in the contemporary world. It is hard to appreciate today that ecologists and economists were so far apart even 2 decades ago that they rarely got together to discuss matters that nowadays are studied by them in collaboration.

My own involvement with the annual Askö meetings arose in the following, somewhat convoluted, way: While writing *The Control of Resources* (1982), I came across the treatise *Ecoscience: Population, Resources, Environment* (1977) by Paul and Anne Ehrlich and John Holdren. That book greatly influenced the way I framed many of the problems I was then studying. But it wasn't until the autumn of 1989 that I first met Paul Ehrlich, at Stanford. Over the next year we become friends, and helped to initiate a seminar that was presided over by Walter Falcon and attended by, among others, Kenneth Arrow, Gretchen Daily, Larry Goulder, Jonathan Roughgarden, David Starrett, and Peter Vitousek.

But our group was an exception: ecologists and economists in the US still viewed one another with deep suspicion. When the Beijer Institute was reconstituted in Stockholm in 1991, its founders had it in mind to bring together ecologists and economists so as to create a *scientific* discipline that was waiting to be established, namely, ecological economics. The choice of Karl-Göran Mäler as the Institute's Director would now be viewed as a stroke of genius but for the fact that it had an obviousness about it. That Ehrlich, Bengt-Owe Jansson, Buzz Holling, Assar Lindbeck, Simon Levin, and I were on the Institute's Scientific Advisory Board in its initial years, and that Charles Perrings was asked to direct one of the Institute's first research programs (on biodiversity) may have helped, but it was Karl-Göran's leadership that enabled the Institute to begin its work with no fuss.

Karl-Göran conceived the idea of an annual meeting of his Advisory Board and invited scholars as an updated version of the annual Solvay conferences that

P. Dasgupta
Faculty of Economics, University of Cambridge, Sidgwick Avenue, Cambridge, CB3 9DD, UK
e-mail: partha.dasgupta@econ.cam.ac.uk

T. Söderqvist et al. (eds.), *Bringing Ecologists and Economists Together*,
DOI 10.1007/978-90-481-9476-6_3, © Springer Science+Business Media B.V. 2011

were held among theoretical physicists in the early years of the last century. Intellectual history tells us that physics was undergoing a revolution in the early 1900s. It tells us too that practicing physicists even then recognized that a revolution was under way. The Solvay conferences were a means for exchanging findings, exploring ways to interpret those findings, and charting future agenda. However, although "Solvay" was the inspiration, "Askö" evolved in a different way. For logistic reasons, Karl-Göran chose to hold the first Askö meeting immediately following the Institute's annual Board meeting in 1993. The Askö meeting lasted one working day, but spread over 2 days, to accommodate the fact the journey from Stockholm to the island takes time. That arrangement has remained pretty much the same.

The island of Askö chose itself for the annual 2-day event because Bengt-Owe had established a marine laboratory there for the University of Stockholm. The logistics of the return journey to Askö from Stockholm aren't straightforward, but Christina Leijonhufvud, Astrid Auraldsson and, in recent years Anna Sundbaum, have ensured that we would not even realize that there were logistic issues.

Although the Solvay conferences were Karl-Göran's inspiration, he asked us to conduct the Askö meetings in a very different form. A theme would be proposed in advance. Although one of us would be asked to introduce it with a set of opening remarks, the discussion that would follow was to be self-consciously a form of a stream of consciousness among a group of people. And that is exactly how the meetings have been conducted since 1993. That we were expected to come up with the beginnings of a paper by the end of the 2 days of discussion must have helped us to concentrate, but it is still puzzling to me how good Askö's record has been. The annual meetings have resulted in notes published in *Science* and a number of other general science journals. In a piece I prepared to celebrate the 10th anniversary of the reconstituted Beijer Institute, I reproduced the notes Carl Folke and I had prepared following the first Askö meeting of 1993. We noted there that the first session had been uneasy, as the two groups vied with each other for time, but that by late afternoon the meeting had settled down into the kind of collegiate encounter we now take for granted at the Beijer Institute. I know of no other event that even faintly resembles Askö. That may be why the Askö meetings are now so greatly admired by ecological economists.

Chapter 4
A Personal View: Saving the World While Having Fun

Paul R. Ehrlich

No serious scholar believes that the human predicament can be solved within the bounds of a single discipline. Be it dealing with climate change, loss of biodiversity, continued population growth, toxification of the planet, a decaying epidemiological environment, or runaway consumption among the rich, the need for input from a diversity of disciplines is self-evident. Ecology and economics are the two disciplines that arguably are the most crucial sources of input into understanding the interactions between Earth's dominant animal, *Homo sapiens*, and the environmental systems that support it. Two decades ago these disciplines seemed to be at loggerheads with one another, and as an ecologist I rarely interacted with economists and assumed their views were expressed by sources such as *The Economist* and the editorial pages of the *Wall Street Journal*. Askö meetings have dramatically changed all that, greatly increased my appreciation of economics, and have moved me to the point where some of my closest colleagues and friends are economists. How did such a miracle happen?

It has its roots in getting to know Partha Dasgupta during the sabbatical year he spent at Stanford. Partha was deeply interested in ecology and fired my interest in economics. In part through that connection, I was asked to serve on the founding Beijer Institute Board. I had my doubts, but they evaporated at my first Askö meeting. There the Gretchen Daily Hypothesis was turned into a law for me. She long ago advised me that if you want scholars from different disciplines to collaborate, you don't just put a couple of them into a room and tell them to do research. Instead you have seminars and workshops at which chocolate and wine are served. That leads to friendships across disciplinary boundaries, that leads to discovering common interests (at least in chocolate and wine!), and that, in turn, leads to the production of good interdisciplinary research.

Askö meetings, which started in 1993, involve assembling a group of first-rate ecologists and economists, sometimes leavened with, say, an ethicist. The meetings

P.R. Ehrlich
Department of Biological Sciences, Center for Conservation Biology, Stanford University, Stanford, CA 94305-5020, USA
e-mail: pre@leland.stanford.edu

T. Söderqvist et al. (eds.), *Bringing Ecologists and Economists Together*,
DOI 10.1007/978-90-481-9476-6_4, © Springer Science+Business Media B.V. 2011

usually start with a nice dinner in Stockholm, followed by a bus-and-boat trip to the beautiful island of Askö in the Stockholm archipelago. There a day and a half is spent discussing a topic of mutual (and world) interest and starting to write a paper about it. At the beginning I was concerned that the time would be spent arguing between ecologists and economists, and that creating a consensus document would prove impossible. My fears were groundless. In a very short time everyone had a basic grasp of the issues from both sides, and friendly disagreements were as frequent within the two disciplines as between them. Progress in getting consensus on the issues was, to me, usually stunningly rapid. After a day of wide-ranging talk, we would get down to roughing out a paper. It would later be completed by chosen (and rotating) leaders and polished in e-mail correspondence. This rapid convergence was, in part, because the problems of dealing with nature's economy and the human economy are so similar in form that many of the approaches of the two disciplines had evolved in parallel. But in large part convergence stemmed from the mutual understandings and friendship that developed as we shared hikes, meals and evening celebrations at which the group told jokes, drank beer and wine, ate chocolate cookies, and sang songs.

The core Askö group quickly transformed itself into a family, and a family with a purpose. We were all convinced that the world's environmental problems were becoming increasingly serious. And we all realized that planning to ameliorate the most pressing of them, from the impacts of population growth, consumption and global heating on ecosystem services, to the persistence of poverty, hunger, and the threat of vast epidemics all require substantial inputs from both ecology and economics. Thus we had common cause and common interests.

To me some of the most interesting and important discussions at Askö concerned the pressing problem of growth in general and overconsumption. While most numerate people now understand that population growth exacerbates environmental problems and will inevitably end, it is still broadly assumed that consumption can increase forever. The United States and the world have been sliding into a recession, in no small part fueled by limited petroleum supplies and the costs of American efforts to secure the oil flows of Iraq by force. But the cures for such economic woes are still seen by politicians and old-time economists mostly as further growth in the physical economy and in the consumption of its products.

At Askö this has been the topic of much discussion, and the general conclusion so far seems to be that while *more* consumption is needed by the poor, the results of pressing for more consumption by the rich would be at best mixed. Uncertainties made firm conclusions on global consumption impossible, but there was general agreement that vigorous public policies are required to support more efficient consumption and investment choices. The Beijer group has not finished wrestling with this supremely important issue, but it has at least presented its dilemmas to the scholarly community.

Among related crucial issues addressed in Askö meetings have been Earth's carrying capacity, valuing natural capital, the relationship between population growth and food security, the resilience of our life-support systems, and the overall problem (repeatedly injected by my friend Simon Levin) of how policy makers might deal with the emergent properties of the global ecology–society complex adaptive system. The importance of the Askö meetings has extended far beyond the small-group discussions and often seminal papers that have flowed from them. Groups of people have been created that are carrying their impact far and wide. The Resilience Network is an outstanding example.

From a personal viewpoint Askö triggered a long-term set of inspiring friendships and collaborations with my Stanford colleagues Ken Arrow and Larry Goulder, and continuing interactions with Partha Dasgupta, now stretching over 2 decades. New friendships with economists such as Tasos Xepapadeas and Aart de Zeeuw have led to further interesting interactions. More recently a high point was at the 2007 meeting of the Association of Environmental and Resource Economists in Europe, which featured an informative debate with John List, dean of economists who actually do field experiments. Best of all, if it were not for Askö I probably never would have become friends with Sara Aniyar, Scott Barrett, Carl Folke, Geoff Heal, AnnMari and Bengt-Owe Jansson, Christina Leijonhufvud, Charles Perrings, Steve Polasky, and many others. And I certainly would not have discovered that Karl-Göran Mäler, the brilliant éminence grise behind the meetings, shared with me a passion for bird watching!

Chapter 5
Building Bridges Between Ecology and Economics

Simon A. Levin

We are facing unprecedented challenges in addressing environmental problems, challenges that raise scientific issues that are both ecological and socioeconomic. Unfortunately, too often, the practitioners of ecology and economics are viewed as being at the two extremes of a dipole, in eternal conflict with one another because of the seemingly incompatible goals of economic growth and environmental protection. Some ecologists, and even a few renegade economists, viewing classical economics as hopelessly inflexible in its single-minded adherence to principles of economic growth, have attempted to create a new discipline of environmental economics, largely insulated from the great body of economists and economic thought. This is, in my view, a terrible shame, and the Askö meetings (not always at Askö, of course) provide the *prima facie* evidence supporting my thesis.

Indeed, ecologists and economists have much of interest in common. It is impossible to address the management of biodiversity without considering the economic importance of that biodiversity, the costs of preservation or the socio-economic forces that guide individual decision-making. On the other hand, sustained economic health is impossible if environmental health is ignored. The linkages between ecological and socio-economic systems are profound and complex; and increasingly, we ignore those linkages at our peril. It is the mutual recognition of the centrality of this principle that has attracted a diverse group of ecologists and economists to the Beijer Institute, and to Askö, and the partnership continues to thrive because of the depth of this commitment. It has worked only because the economists involved carry unassailable credentials and respect within the economics community, and because they hold to the principle that the problems addressed are of surpassing importance to humanity and can only be solved through cooperation with ecologists. Similarly, it has worked only because the ecologists involved carry unassailable credentials within the ecological and the conservation communities, share the concern for the problems being addressed, and believe that solutions are only possible through partnerships with economists. It has only

S.A. Levin
Department of Ecology and Evolutionary Biology,
Princeton University, Princeton, NJ 08544, USA
e-mail: slevin@princeton.edu

T. Söderqvist et al. (eds.), *Bringing Ecologists and Economists Together*,
DOI 10.1007/978-90-481-9476-6_5,

worked because the principals, over a period approaching 2 decades, have developed mutual trust and admiration, common language and friendship. It has only worked because of the inspired leadership of Karl-Göran Mäler, who found the right people, the right themes and the right venues to create progress in addressing some of the hardest issues facing humanity.

It is the applied challenges that have brought these diverse communities together, but I think that progress would have been much more difficult, and progress short-lived, were it not for the early recognition among all involved that their sciences were not so different, and that there are exciting theoretical questions that are common to both disciplines. Ecological systems are, after all, special cases of economic systems, in which the currencies are energy or other limiting resources, and in which competition for those resources involves the interplay among individuals carrying out their own selfish agendas. Or, viewed symmetrically, economic systems are special cases of ecological systems, in which evolutionary and other forces build communities of cooperators and competitors, and of parasites and predators. Ecological and socio-economic networks are alike as complex adaptive systems, in which heterogeneous assemblages of individual agents interact on multiple scales, and in which evolutionary forces (genetic and otherwise) alter the composition and topologies of those networks in ways that feed back to affect the dynamics on local scales. The interplay among the processes operating on multiple scales produces the complexity that characterizes these systems, and raises a range of computational and intellectual challenges that benefit from the cross-fertilization among scientists of diverse disciplines. The commonality of these challenges is not restricted to ecology and economics of course, but extends beyond the social sciences to physics and cosmology, to chemistry and molecular biology. In all of these, the challenge of building bridges between the microscopic and the macroscopic viewpoints is shorthand for the need to relate phenomena across scales, and to understand how individual actions translate into consequences for collectives. Such issues have been at the core of the Askö meetings.

Any list of the central problems in the theory of complex adaptive systems would find resonance in the themes discussed at Askö. Can systems flip from one stable configuration to another, for example lakes from oligotrophy to eutrophy, or economic systems from boom to bust? Are there signals that herald fragility and loss of resilience or robustness? How has evolution shaped the nature of systems, and the ways that individuals make their decisions and discount the future? Under what conditions can cooperation arise, and how can we build on that understanding to achieve a sustainable future?

But Askö has balanced such core theoretical questions with the practical as well. How do we measure the value of services ecosystems provide humanity, and create a basis for green accounting? How can the externalities associated with life in a global commons be internalized in the accounting scheme? How do we assess the human carrying capacity of the globe, the appropriate balance between consumption and environmental sustainability, or the fragility of ecotone ecosystems like the Arctic? What are the likely costs and benefits of free trade for natural and managed ecosystems?

Askö has explored all of these issues and more, and the papers reprinted in this volume underscore the brilliance of Karl-Göran Mäler's vision, and the success of his implementation strategy. The early realizations of Askö were remarkably successful, given the need for the players to learn to speak to each other, to trust each other, and to anticipate each other. Those critical steps were achieved surprisingly quickly, and the later meetings have been like the later chapters in an unfolding story, building on the earlier chapters while bringing in new and essential elements. The success of the model is always surprising to newcomers. How, after all, can one achieve meaningful consensus in less than 2 days in the absence of a great deal of preparatory work? But it has worked, and worked brilliantly, and the contributions chronicled in this volume represent a major step forward in building a true ecological economics.

Chapter 6
Guidelines for the Future?

AnnMari Jansson and Bengt-Owe Jansson

6.1 How It All Started

We clearly remember the first time the idea of the Askö meetings emerged as a glimt in the eyes of Karl-Göran Mäler a sunny June afternoon 1989 at the Askö Laboratory in the Stockholm Archipelago. He had just returned from a birdwatching stroll on the island with a colleague. When wiping the sweat off his face he looked around and said: "This would be the perfect place for seminars: eating, drinking, birdwatching while discussing ecology and economy". Actually the place was well motivated also for other reasons.

During the 1970s, the station hosted the research project "Dynamics and energy flow of the Baltic Ecosystem" financed by the Swedish Natural Research Council. Spending 1971–1972 with Howard T. Odum in Florida, we were learning systems ecology and modeling and became fascinated of his ideas of the energy basis of nature as a life-supporting system of man, its hierarchical composition, the importance of feedback loops, species diversity as information. The effects of man's use of fossil fuel, competing with and multiplying nature's processes (*potatoes made by oil*) became a real eye-opener. This was quite new to us! Back home we applied his ideas in writing and field-working to the total Baltic Sea. This inspired also the scientists in the other Baltic countries to treat the Baltic Sea as a total ecosystem. Odum's theories of including also the flow of money in the systems models were applied by AnnMari to a clear-cut part of Sweden, the island of Gotland in the Baltic. During an 8-year project starting in 1974, the flows of energy and money were mapped and analyzed with an advisory board, appointed by the Swedish research councils. One member of the board was Karl-Göran Mäler. The activities were received by the community with more courtesy than enthusiasm. To force the scientific community to take stand AnnMari initiated an international workshop in 1982: *Integration of Economy and Ecology – an*

A. Sundbaum
The Beijer Institute of Ecological Economics, the Royal Swedish Academy of Sciences, Box 50005, SE-104 05, Stockholm, Sweden
e-mail: anna.sundbaum@gmail.com

T. Söderqvist et al. (eds.), *Bringing Ecologists and Economists Together*,
DOI 10.1007/978-90-481-9476-6_6, © Springer Science+Business Media B.V. 2011

Outlook for the Eighties, financed by the Wallenberg Foundation for International Cooperation in Science. One member of the planning board was Karl-Göran Mäler and among the invited international scientists were Robert Costanza, Frank Golley, David Pimentel, Herman Daly, Ralph D'Arge, Partha Dasgupta, Anthony Fisher, Allan Kneese and Howard Odum. The lively (to say the least) discussions, including clashes between more or less well-grounded arguments, are most interesting to compare with the more "well-temperated" Askö meetings. The continued contacts and collaboration between AnnMari, Bob Costanza and Herman Daly led to the foundation of the International Society for Ecological Economics in 1989.

6.2 The Conceptual Evolution of the Discussions

Most of the meetings are documented in already printed, scientific articles but we cannot refrain from picking for us tasty raisins from some of them, to show how economic and ecological theory and reality actually can be reconciled and translated to practical advice for sustainable resource use.

In the first meeting in 1993 the whole arsenal from the battles around the new discipline "ecological economics" was exercised by the more or less already "conditioned" participants, including two of the founders (Bob Costanza and AnnMari). Comparison of economic and ecological models generated discussions on the dependence of the natural resource base. The far from stable ecosystem with its inherent complexity, thresholds and "surprises" motivated a strong recommendation of including "the precautionary principle" in decision-making. The common market was unanimously rejected as a suitable allocator of resources.

"Economic growth" was the controversial concept in 1994. It generated a stimulating, constructive and colorful discussion, reported in *Science*. Here our personal struggles on the Swedish arena to get an understanding for the importance of the ecosystem got international verification. Present economic policy regards the planet as capable of supporting indefinite economic growth, ignoring accumulation of substances, feedback effects, irreversible changes like desertification and groundwater depletion. The content of growth is what matters, and the composition of inputs and outputs. The limit of carrying capacity of the globe is set by the properties of the natural system like complexity, interactions, resilience, multistable equilibria. The economic institutions are important, they have to ensure the scale of economic activity and make sure that the social costs are internalized.

The meeting in 1995 focused on the concept of resilience, *sensu* Holling, in natural and economic systems. It is non-linear and adaptive and implies complex and far-from-equilibrium dynamics. Small changes lead to unexpected events. The hierarchical feedback mechanisms generate multistable equilibria, broken by Schumpeter's "creative destruction", like fires in forest systems or the shift in the economic system in the former Soviet Union. Resilience is created through feedback mechanisms, competition between organisms in natural systems, between market participants in economic systems. The ignorance of the economic system of the resilience of ecosystems has been clearly demonstrated also in the Swedish public discussions of climate change. We are very happy for the advice to direct the market forces towards sustainable decisions by changing institutions, norms and goals to tighten the crucial feedback loops.

The problem of population growth and food production, addressed at the 1997 meeting, generated strong criticism for the present measures of welfare. Conventional indicators of the standard of living refer to commodity production and not to the natural resource base. GNP, market price, life expectancy, etc., have no connection to the strong couplings between population growth, increased material consumption and the state of the natural resource base. This ignorance has resulted in destruction of genetic resources, erosion of the ecosystem services, health impact of toxic fertilizers and pesticides, extreme climatic events. Globalization increasingly affects local food availability, moving local food systems into a more non-linear domain. Instead we should invest more in local plant material, including dissemination of new technique and mapping decreasing populations of wild crops. In Sweden, for example, we strongly advocate a change to a more ecological agriculture. Specialized and now geographically isolated grain and meat producing farms, respectively, should be merged to secure recycling of nutrients and local production should generate local consumption to decrease waste of energy and environmental effects of transportation.

Focusing on nature as the life support system of man that Odum so hard argued for, the discussions in 1998 treated the ecosystem services connected to food production and population growth. Through the previous meetings we had at last reached the point of "Put a Price on the Environment" (to cite the 1968 book of the Swedish economist Erik Dahmén). Now a further concrete step was taken: The ecosystems of the world were simply declared as capital assets and targets for conservation. We have to characterize the major production functions, map the sources and consumers and define the critical points and interdependencies in the supply of services. To amalgamate ecology and economy in decision-making we must identify possible alternatives, impacts and consequences of suggested alternatives. How much are people willing to sacrifice to get expected outcomes? How can individual values be integrated to social values?

The nonlinear character of the natural systems, including multistable equilibria and "surprises", makes "uncertainty" to a basic concept in all forms of resource use, and probably to the greatest obstacle in pursuing people to actively work for a sustainable society. "Uncertainty" was the theme of the 2002 meeting, which, possibly due to the "conditioning" effects of the previous meetings, offered a most constructive discussion, summarized in the very readable article, published in *Ambio*. The reasons for the difficulties in understanding between scientists and policy makers were dissected. The strong scientific tradition of precise evidence delays information to policymakers; scientists are reluctant to quantify uncertainty, pleading insufficient data and questioning of data quality displaces a fundamental debate on values. Application of two well-established principles is recommended: "Expected utility" for making decisions and "Bayesian updating" for revising the decisions at a later date with more available information. Scientists should report their available results with whatever level of confidence they can provide. "Surprises" represent a case of "functional uncertainty". Recent examples are the crash of the cod fisheries at Georges Bank and in the Baltic Sea, where overfishing of the top predator generates a "cascade" of changes in the food chain and the total ecosystem. In public decision making the likelihood of a "bad surprise" must be traded off against the lost benefit, both social and economic. General and practical lessons for coping with uncertainty include: Politicians must be willing to risk votes in order to follow socially and scientifically recommended pathways; institutions must be made flexible in order to cope with surprises; society must produce educated citizens.

6.3 Beijer on Askö[1]

Looking back it has been both an honor and a great pleasure to participate in those meetings with internationally outstanding members of the concerned disciplines. It has been interesting to follow, how the original rather divergent views from the respective economic and ecological "windows" successively merged to one large and broad view of the real world of which the basic concepts of sustainability have got their true and fuller dimensions. After decades of pleading with environmental data we have at last got basic economic arguments for using ecosystems principles in resource exploitation!

[1]A suitable Swedish translation: "Askö and Beijer – finfina grejer".

Part III
The Askö Papers

Chapter 7
Askö 1993: Ecologists and Economists Can Find Common Ground*

Carl Folke

In recent years there has been an encouraging trend of increased interdisciplinary collaboration between ecologists and economists. But, as a general rule, the writings on environmental matters by ecologists and by economists are different in tenor and message. The economists on the whole appear to be more optimistic when regarding the condition of the human environment; in particular, the economists tend to believe that technological innovations can be relied upon to solve environmental problems, while ecologists are less inclined to trust technology to cure or bypass problems.

Do the different attitudes of ecologists and economists reflect different value systems, that is, do members of these professions hold different world views? In short, are ecologists nature lovers, while economists are materialists? And if so, is there self-selection of these types into the two professions? In a discussion in Askö, Sweden, prominent members of the two professions[1] concluded that although members of the two groups may on average come from different backgrounds, there is no evidence that they hold substantially different values. Ecologists have been known to enjoy high consumption levels, while economists have been known to love nature walks. More likely, it was felt, the professional differences were generated elsewhere.

*This article was published in *BioScience*, Vol. 45, Folke et al., Ecologists and economists can find common ground, 283–284, Copyright AIBS (1995).

[1]The group comprised ecologists R. Costanza, Maryland International Institute for Ecological Economics; P. R. Ehrlich, Stanford University; C. Folke, Beijer International Institute of Ecological Economics and Stockholm University; C. S. Holling, University of Florida; A.-M. Jansson and B.-O. Jansson, Stockholm University; and J. Roughgarden, Stanford University; and economists P. Dasgupta, University of Cambridge; G. M. Heal, Columbia University; K.-G. Mäler, Beijer International Institute of Ecological Economics and Stockholm School of Economics; C. Perrings, University of York; and D. A. Starrett, Stanford University. The meeting, sponsored by Beijer International Institute of Ecological Economics (which is part of the Royal Swedish Academy of Sciences), was held at the Askö Laboratory of the Stockholm Centre for Marine Research, Askö, Sweden, 5–7 September 1993.

C. Folke
Department of Systems Ecology, Stockholm University and Beijer Institute of Ecological Economics, the Royal Swedish Academy of Sciences, Box 50005, SE-104 05, Stockholm, Sweden

T. Söderqvist et al. (eds.), *Bringing Ecologists and Economists Together*,
DOI 10.1007/978-90-481-9476-6_7, © Springer Science+Business Media B.V. 2011

Neither discipline has sufficiently considered the inputs of the other discipline, according to the discussion participants. Models in each field are constructed for specific ends. Not all economic models need include environmental variables, any more than there need be an economic element in all ecological models. However, the economists felt that too many economic models ignore the environmental resource base of material production (e.g., deterioration of mangrove ecosystems into shrimp aquaculture) and the consequences of that production for critical environmental systems employed as sinks (e.g., the atmosphere as a sink for carbon dioxide emitted in the process of burning fossil fuels in rich countries).

Those models continue to postulate unlimited growth in population, unlimited growth of the physical economy by means of capital accumulation and substitution, improved organization, and technological progress. The nature of these models, the economists suggested, may well adversely affect their profession's perception of the natural world. They noted too that, as a profession, economists tend to overly stress the ability of markets to allocate resources efficiently. For example, in view of the lack of well-defined property rights in most environmental resources and sinks, failure to consider externalities (which exist when prices do not reflect true social costs) is pervasive. These externalities are not given the prominence they require and in textbooks are still regarded as aberrations.

Another long-standing weakness of the economist's modeling of the environment is that, on the whole, it ignores possible threshold effects, a central concern of ecologists. The economists explained that if threshold effects are significant, the price mechanisms on which economists rely cannot perform well.

The ecologists in turn feel that, when searching for solutions to environmental problems, members of their discipline all too often fail to take advantage of the knowledge of economists and other social scientists, in particular regarding the importance of markets for allocating environmental resources. Frequently ecologists do not bring critical scrutiny to bear on environmentalists who may support central command and control measures when a market mechanism may be more efficient, and vice versa. Ecologists often do not appreciate the underlying economic causes or other driving forces of environmental problems or the indirect effects of remedial measures proposed.

Better communication is needed between ecologists and environmentalists, as well as between ecologists and economists. The ecologists at the Askö meeting also felt that on issues such as the relationship between complexity and stability, they have not sufficiently informed environmentalists of the current state of their science. As a result, environmentalists may use as slogans some notions that are outdated. Ecologists have also tried to move quickly from an understanding of small-scale case studies to predictions about large-scale systems. They may therefore appear more certain than is justified about what they know about the behavior of large systems. In view of the limited ecological understanding, the ecologists at the meeting recommended a precautionary approach to treatment of large systems out of a fear of the consequences of their possible collapse.

The ecologists and economists agreed that there are often substantial conceptual differences between their population, resource, and environmental models due to ignorance of each other's fields. It is bad science when economists build models that are oblivious of ecological knowledge (such as the limited substitution possibilities among resources), and bad science can lead to bad policy and faulty management. By the same token, if ecological models ignore the ways in which economic institutions operate, they too can have unsatisfactory consequences.

The ecologists expressed concern that most economists continue to view as an unalloyed good the growth of the gross national product in rich nations with high levels of wasteful consumption. They emphasized that it was necessary for the material economies of poor nations to grow but that this growth should be balanced by decreasing throughput in rich nations – something they claimed could be achieved with an improvement in the quality of life. They argued that increasing the scale of the global human enterprise (that is, the product of population size and per capita consumption) is a recipe for environmental disaster. Therefore, they are alarmed that economic analyses of the global economy often do not capture the critical relationship of the scale of the human economy to the scale of the ecosystems that support it.

The economists agreed that the gross national product is not an ideal measure of human welfare and that it is all too often misinterpreted in the press and by politicians. They noted, however, that economists have in recent years put considerable effort into devising improved measures. The economists shared the ecologists' concern on the importance of global-scale issues, because they agree that the world's natural capital is increasingly becoming scarce. The two groups also agreed on the need for a careful reconsideration of where and how economic growth and shrinkage should be pursued. An overall conclusion was that the ecologists know more than enough to alert a risk–adverse society to directions in which it should move in order to avoid serious environmental catastrophes and that economists' expertise would be critical in designing mechanisms that would encourage that movement.

Chapter 8
Askö 1993: Commentary by Ann P. Kinzig

Ann P. Kinzig

The first Askö meeting – the beginning of a series of productive meetings between leading ecologists and economists – concerned itself with, among other things, whether the different values systems of economists and ecologists might preclude successful collaboration. Perhaps there was reason to be suspicious – a reading of the economic literature might suggest that economists viewed nature as much more resilient and fungible than ecologists believed it to be; a similar reading of the ecological literature might reveal deep misgivings towards the capacity of any modern human institution to properly manage natural resources.

And yet there were many other reasons to hope that fruitful cross-fertilization between the two disciplines could occur. While it may startle many ecologists to hear so, economists are their closest natural intellectual allies. Both disciplines are interested in the attributes that confer success on individuals or agents (e.g., populations, firms) in different and sometimes rapidly changing contexts. Both assert that certain system-level properties (e.g., nutrient retention, efficiency) will emerge as the result of the fundamentally selfish behaviors of these individuals or agents. Many of the economic models used to examine emergent system-level properties in the marketplace could, with little modification, be effectively applied to ecological systems; many of the strategies for survival revealed by ecologists find their analog in human marketplace behavior. Pursuit of an understanding of this dance – small-scale interactions leading to large-scale emergent properties that themselves constrain the small-scale interactions – is one of the most important fundamental intellectual pursuits in science today, and one that is conducted with equal vigor by both ecologists and economists.

As the first Askö paper noted, there are pressing practical, as well as intellectual, reasons to forge an alliance between ecologists and economists. Ecological insights into diversity–functioning relationships, and functional redundancy and diversity, need to inform economic assumptions about the fungibility of natural resources.

A.P. Kinzig
School of Life Sciences, Arizona State University, P.O. Box 874501, Tempe, AZ 85287-4501, USA
e-mail: kinzig@asu.edu

T. Söderqvist et al. (eds.), *Bringing Ecologists and Economists Together*,
DOI 10.1007/978-90-481-9476-6_8,

A growing awareness among ecologists – present in the early 1990s but even more profound today – of the non-linearities and thresholds operating in ecological systems call into question many models of the environment used by economists which, with their assumptions of smooth response, may conclude that certain levels of resource exploitation are reasonable when in fact they may bring the resource base, and thus human well being, perilously close to collapse.

At the same time, the policies or strategies that ecologists might offer to better manage the natural systems they so love are naïve at best, and counterproductive at worst, when uninformed by fundamental economic insights. Unbeknownst to too many ecologists, economics is not a discipline dedicated to justifying rabid exploitation of any and all resources in the pursuit of individual wealth, but a discipline that reveals strategies for structuring the reasonable allocation of resources given the articulated values of society towards those resources. It is a positive and not a normative science. Ignoring what it has to offer means proposing solutions to resource–management problems that, while potentially pleasing from the perspective of the individual (and likely well fed and otherwise economically comfortable) ecologist, would be far from optimal from the perspective of a society navigating complex trade-offs. The science of ecology tells us nothing about navigating these trade-offs; the science of economics takes that challenge as its starting point.

The exchange of purely intellectual insights – about emergent properties, non-linearities, alternative management strategies – can afford a certain laziness. There are reasons for conservatism in science, and to the extent that science is about satisfying a deep and persistent human curiosity, we can afford to take our time. Our curiosity will still be there tomorrow. But there is more urgency when one considers what the failure of a marriage between ecology and economics might mean on a practical level. At best, it would mean that the world's people and ecosystems would be marginally worse off than they might otherwise have been. At worst, and potentially more likely, it would mean persistent and profoundly debilitating impoverishment for at least a fraction of the world's people, and the irreversible loss of countless irreplaceable ecological treasures.

The Beijer Institute took a first step in that courtship when it held its first Askö meeting. It has continued to build that relationship through subsequent meetings, and the results of those held in 1993–2002 are highlighted in this volume. Beijer has done more than anyone could expect of a relatively small institute. But it is not enough; implemented and sensible approaches to natural-resource management still elude us. It is my hope that others reading this volume come to recognize the value of the marriage between ecologists and economists, and find many other ways and places to achieve that union.

Chapter 9
Askö 1993: Commentary by Charles Perrings

Charles Perrings

This marked the beginning of the Askö experiment – an annual meeting involving a group of variable composition, but always including a mix of economists and ecologists with a direct interest in human interactions with the natural environment. The aim of the first meeting was to identify the intersection not just in the foci of the two disciplines, but also in the value systems of people working in each area. Today, the second concern seems a little naive. Why would one expect people with a professional concern for the environment, but working from different disciplinary perspectives, to have fundamentally different value systems? Yet there was, at the time, a suspicion on the part of the ecologists brought into the Beijer Institute's orbit that economics was all about making money, and that economists shared the acquisitive and materialistic instincts of the people whose behavior they studied. The fact that many of the people who have participated in the Askö meetings over the years do share similar values has certainly helped the debate, but it is not the important part of the 'common ground' that Folke writes about.

The important 'common ground' in fact lies in the recognition that the production and consumption activities of economic agents affect and are affected by a set of natural processes. There are feedbacks between the economy and its environment that drive the dynamics of the whole system. In other words, both economists and ecologists recognized that the systems they studied were inextricably linked, that the linkages matter, and that the tools of each discipline separately can give only partial purchase on the world. More importantly, they recognized that failure to take this into account could lead to misleading conclusions. As Folke put it: 'It is bad science when economists build models that are oblivious of ecological knowledge (such as the limited substitution possibilities among resources), and bad science can lead to bad policy and faulty management'.

Many of the assumptions conventionally made by economists, even those working on the environment, were questioned in the first Askö meeting. Aside from the degree of substitutability between produced and natural capital, doubt was

C. Perrings
School of Life Sciences, Arizona State University, P.O. Box 874501, Tempe, AZ 85287-4501, USA
e-mail: charles.perrings@asu.edu

T. Söderqvist et al. (eds.), *Bringing Ecologists and Economists Together*,
DOI 10.1007/978-90-481-9476-6_9, © Springer Science+Business Media B.V. 2011

raised about the assumption of the convexity of production sets. It was argued that many of the processes of the natural environment exhibit all the usual properties of complex systems, including sensitivity to initial conditions, path dependence, non-linearity and so on. However, these properties were seldom treated by economists. It was also argued that scale issues – important because of the limited carrying capacity of ecosystems – were systematically neglected. Threshold effects were all but ignored.

On the other side, economists were concerned by the extent to which ecologists persisted in focusing on 'natural systems', ignoring the structuring effect of economic activities. This is less prominent in Folke's paper, but economists were also concerned at the incompatibility of two interpretations of conservation. By one interpretation, conservation is the preservation of natural systems. By another it is the management of the coupled system with a view to assuring sustainable use. Understanding the pervasiveness of human influence on the natural environment, and the critical importance of human decisions for conservation was therefore part of the common ground between economics and ecology. The implication of this is that the regulatory and market-based tools of economics are an integral part of conservation.

Since the first Askö meeting, considerable attention has been paid in the economic literature to the problems posed by non-convexities in the coupled system (e.g. Perrings 1998; Brock et al. 2002; Mäler et al. 2003; Brock and Starrett 2003; Brock and Xepapadeas 2004; Dasgupta and Mäler 2004), and to the ecological implications of threshold effects induced by human exploitation of natural environments (e.g. Carpenter et al. 1999; Levin et al. 1998; Levin 1999; Gunderson and Holling 2002; Carpenter and Brock 2004; Walker et al. 2004). In some cases this literature was a direct consequence of the Askö experience, in others not. In all cases, however, the authors have benefited from the opportunity to exchange views in an open unstructured setting where the only crime is disengagement.

References

Brock, W. A., Mäler, K.-G., & Perrings, C. (2002). Resilience and sustainability: The economic analysis of non-linear dynamic systems. In L. H. Gunderson & C. S. Holling (Eds.), *Panarchy: Understanding transformations in systems of humans and nature* (pp. 261–291). Washington, DC: Island Press.

Brock, W. A., & Starrett, D. (2003). Nonconvexities in ecological management problems. *Environmental and Resource Economics, 26*(4), 575–624.

Brock, W. A., & Xepapadeas, A. (2004). Management of interacting species: Regulation under nonlinearities and hysteresis. *Resource and Energy Economics, 26*, 137–156.

Carpenter, S. R., & Brock, W. A. (2004). Spatial complexity, resilience and policy diversity: Fishing on lake-rich landscapes. *Ecology and Society, 9*(1), 8. [online] URL: http://www.ecologyandsociety.org/vol9/iss1/art8.

Carpenter, S. R., Ludwig, D., & Brock, W. A. (1999). Management of eutrophication for lakes subject to potentially irreversible change. *Ecological Applications, 9*(3), 751–771.

Dasgupta, P., & Mäler, K.-G. (2004). *The economics of non-convex ecosystems*. Dordrecht: Kluwer.

Gunderson, L. H., & Holling, C. S. (Eds.). (2002). *Panarchy: Understanding transformations in systems of humans and nature*. Washington, DC: Island Press.
Levin, S. A. (1999). *Fragile dominion: Complexity and the commons*. Reading, MA: Perseus Books.
Levin, S., Barrett, S., Aniyar. S., Baumol, W., Bliss, C., Bolin, B., et al. (1998). Resilience in natural and socioeconomic system. *Environment and Development Economics*, 3, 222–235.
Mäler, K.-G., Xepapadeas, A., & de Zeeuw, A. (2003). The economics of shallow lakes. *Environmental and Resource Economics, 26*(4), 603–624.
Perrings, C. (1998). Resilience in the dynamics of economy–environment systems. *Environmental and Resource Economics, 11*(3–4), 503–520.
Walker, B., Holling, C. S., Carpenter, S. R., & Kinzig, A. (2004). Resilience, adaptability and transformability in social–ecological systems. *Ecology and Society, 9*(2), 5. [online] URL: http://www.ecologyandsociety.org/vol9/iss2/art5/.

Chapter 10
Askö 1994: Economic Growth, Carrying Capacity, and the Environment*

Kenneth Arrow, Bert Bolin, Robert Costanza, Partha Dasgupta, Carl Folke, C.S. Holling, Bengt-Owe Jansson, Simon Levin, Karl-Göran Mäler, Charles Perrings, and David Pimentel

---

*This article was published in *Science*, Vol. 268, Arrow et al., Economic growth, carrying capacity, and the environment, 520–521, Copyright AAAS (1995).

K. Arrow
Department of Economics, Stanford University,
Stanford, CA 94305, USA

B. Bolin
Department of Meteorology, University of Stockholm, 10691 Stockholm, Sweden

R. Costanza
Maryland International Institute for Ecological Economics, University of Maryland, Box 38, Solomons, MD 20688, USA

P. Dasgupta
Faculty of Economics, Cambridge University, Cambridge CB3 9DD, UK

C. Folke
Beijer Institute of Ecological Economics, the Royal Swedish Academy of Sciences, Box 50005, SE-10405 Stockholm, Sweden

C.S. Holling
Department of Zoology, University of Florida, Gainesville, FL 32611, USA

B.-O. Jansson
Department of Systems Ecology, University of Stockholm, S-10691 Stockholm, Sweden

S. Levin
Department of Ecology and Evolutionary Biology, Princeton University,
Princeton, NJ 08544, USA

K.-G. Mäler
Beijer Institute of Ecological Economics, the Royal Swedish Academy of Sciences, Box 50005, SE-10405 Stockholm, Sweden

C. Perrings
Department of Environmental Economics and Environmental Management, University of York, York Y01 5DD, UK

D. Pimentel
Department of Entomology and Section of Ecology and Systematics, Cornell University, Ithaca, NY 14853, USA

T. Söderqvist et al. (eds.), *Bringing Ecologists and Economists Together*,
DOI 10.1007/978-90-481-9476-6_10, © Springer Science+Business Media B.V. 2011

National and international economic policy has usually ignored the environment. In areas where the environment is beginning to impinge on policy, as in the General Agreement on Tariffs and Trade (GATT) and the North American Free Trade Agreement (NAFTA), it remains a tangential concern, and the presumption is often made that economic growth and economic liberalization (including the liberalization of international trade) are, in some sense, good for the environment. This notion has meant that economy-wide policy reforms designed to promote growth and liberalization have been encouraged with little regard to their environmental consequences, presumably on the assumption that these consequences would either take care of themselves or could be dealt with separately.

In this article we discuss the relation between economic growth and environmental quality, and the link between economic activity and the carrying capacity and resilience of the environment.[1]

10.1 Economic Growth, Institutions, and the Environment

The general proposition that economic growth is good for the environment has been justified by the claim that there exists an empirical relation between per capita income and some measures of environmental quality. It has been observed that as income goes up there is increasing environmental degradation up to a point, after which environmental quality improves. (The relation has an "inverted-U" shape.)

One explanation of this finding is that people in poor countries cannot afford to emphasize amenities over material well-being. Consequently, in the earlier stages of economic development, increased pollution is regarded as an acceptable side effect of economic growth. However, when a country has attained a sufficiently high standard of living, people give greater attention to environmental amenities. This leads to environmental legislation, new institutions for the protection of the environment, and so forth.

The above argument does not, however, pertain to the environmental resource basis of material well-being, a matter we shall return to subsequently.

So far the inverted U-shaped curve has been shown to apply to a selected set of pollutants only (Shafik and Bandyopadhay 1992).[2] However, because it is consistent with the notion that people spend proportionately more on environmental quality as their income rises, economists have conjectured that the curve applies to

[1]This is a report of the Second Askö Meeting, held 31 August to 2 September 1994, in the archipelago outside Stockholm, Sweden. The meeting was organized by the Beijer International Institute of Ecological Economics, Royal Swedish Academy of Sciences, Box 50005, SE-10405 Stockholm, Sweden. The aim of the meeting was to establish a substantive dialogue among a small group of ecologists and economists to gauge whether an interdisciplinary consensus exists on the issue of economic growth, carrying capacity, and the environment and to determine what could be said about the joint development of economic and environmental policy.

[2]Phenomena for which the relation holds most clearly include poor sanitation, impure water supplies, suspended particulates, SO_2, NO_x and CO (Grossman and Kreuger 1993).

environmental quality generally (Beckerman 1992). But it is important to be clear about the conclusions that can be drawn from these empirical findings. While they do indicate that economic growth may be associated with improvements in some environmental indicators, they imply neither that economic growth is sufficient to induce environmental improvement in general, nor that the environmental effects of growth may be ignored, nor, indeed, that the Earth's resource base is capable of supporting indefinite economic growth. In fact, if this base were to be irreversibly degraded, economic activity itself could be at risk (Jansson et al. 1994).

There are other reasons for caution in interpreting these inverted U-shaped curves. First, the relation has been shown to be valid for pollutants involving local short-term costs (for example sulfur, particulates, and fecal coliforms), not for the accumulation of stocks of waste or for pollutants involving long-term and more dispersed costs (such as CO_2), which are often increasing functions of income.[3]

Second, the inverted-U relations have been uncovered for emissions of pollutants, not resource stocks. The relation is less likely to hold wherever the feedback effects of resource stocks are significant, such as those involving soil and its cover, forests, and other ecosystems.

Third, the inverted-U curves, as they have been estimated, say nothing about the system-wide consequences of emission reductions. For example, reductions in one pollutant in one country may involve increases in other pollutants in the same country or transfers of pollutants to other countries (Lopez 1994; Selden and Song 1994; Anderson and Blackhurst 1992).

And fourth, in most cases where emissions have declined with rising income, the reductions have been due to local institutional reforms, such as environmental legislation and market-based incentives to reduce environmental impacts. But such reforms often ignore international and intergenerational consequences. Where the environmental costs of economic activity are borne by the poor, by future generations, or by other countries, the incentives to correct the problem are likely to be weak. The environmental consequences of growing economic activity may, accordingly, be very mixed.

The solution to environmental degradation lies in such institutional reforms as would compel private users of environmental resources to take account of the social costs of their actions.[4] The inverted-U relation is evidence that this has happened in

[3]Shafik and Bandyopadhay (1992) find that CO_2 emissions are an increasing function of per capita income. Moomaw and Tullis (1994) show that while cross-sectionally CO_2 emissions per capita rise with income per capita, the experience in individual countries over time is highly variable, depending on the structure of individual economies. They conclude that there are many different development paths, some of which can de-link economic growth from CO_2 emissions, but adequate institutions are required and it is by no means an automatic consequence of growth.

[4]A related issue is the measurement of economic growth at the macro level. The conventional measure, GNP, is far from adequate as a measure of true economic performance. We need a more comprehensive index that includes the flow of environmental services as well as the value of net changes in the stocks of natural capital so that the true social costs of growth can be internalized. With this improved index, many of the apparent conflicts between "growth" and the environment would disappear because environmental degradation would be subtracted from the index. See, for example, Dasgupta and Mäler (1991).

some cases. It does not constitute evidence that it will happen in all cases or that it will happen in time to avert the important and irreversible global consequences of growth.

10.2 Carrying Capacity and Ecosystem Resilience

The environmental resource base upon which all economic activity ultimately depends includes ecological systems that produce a wide variety of services. This resource base is finite. Furthermore, imprudent use of the environmental resource base may irreversibly reduce the capacity for generating material production in the future. All of this implies that there are limits to the carrying capacity of the planet. It is, of course, possible that improvements in the management of resource systems, accompanied by resource-conserving structural changes in the economy, would enable economic and population growth to take place despite the finiteness of the environmental resource base, at least for some period of time. However, for that to be even conceivable, signals that effectively reflect increasing scarcities of the resource base need to be generated within the economic system.

Carrying capacities in nature are not fixed, static, or simple relations. They are contingent on technology, preferences, and the structure of production and consumption. They are also contingent on the ever-changing state of interactions between the physical and biotic environment. A single number for human carrying capacity would be meaningless because the consequences of both human innovation and biological evolution are inherently unknowable. Nevertheless, a general index of the current scale or intensity of the human economy in relation to that of the biosphere is still useful. For example, Vitousek et al. (1986) calculated that the total net terrestrial primary production of the biosphere currently being appropriated for human consumption is around 40%. This does put the scale of the human presence on the planet in perspective.

A more useful index of environmental sustainability is ecosystem resilience. One way of thinking about resilience is to focus on ecosystem dynamics where there are multiple (locally) stable equilibria.[5] Resilience in this sense is a measure of the magnitude of disturbances that can be absorbed before a system centered on one locally stable equilibrium flips to another (Holling 1973). Economic activities are sustainable only if the life-support ecosystems on which they depend are resilient. Even though ecological resilience is difficult to measure and even though it varies from system to system and from one kind of disturbance to another, it may be possible to identify indicators and early-warning signals of environmental stress. For example, the diversity of organisms or the heterogeneity of ecological functions have been suggested as signals of ecosystem resilience. But ultimately, the resilience of systems may only be

[5]Another way is to focus on ecosystem dynamics in the neighborhood of an equilibrium. The object here is to measure the resistance of a system to disturbances and to determine the speed with which it returns to equilibrium. See Pimm (1984).

tested by intelligently perturbing them and observing the response with what has been called "adaptive management".[6]

The loss of ecosystem resilience is potentially important for at least three reasons. First, the discontinuous change in ecosystem functions as the system flips from one equilibrium to another could be associated with a sudden loss of biological productivity, and so to a reduced capacity to support human life. Second, it may imply an irreversible change in the set of options open both to present and future generations (examples include soil erosion, depletion of groundwater reservoirs, desertification, and loss of biodiversity). Third, discontinuous and irreversible changes from familiar to unfamiliar states increase the uncertainties associated with the environmental effects of economic activities.

If human activities are to be sustainable, we need to ensure that the ecological systems on which our economies depend are resilient. The problem involved in devising environmental policies is to ensure that resilience is maintained, even though the limits on the nature and scale of economic activities thus required are necessarily uncertain.

10.3 Economic Growth and Environmental Policy

We conclude that economic liberalization and other policies that promote gross national product growth are not substitutes for environmental policy. On the contrary, it may well be desirable that they are accompanied by stricter policy reforms. Of particular importance is the need for reforms that would improve the signals that are received by resource users. Environmental damages, including loss of ecological resilience, often occur abruptly. They are frequently not reversible. But abrupt changes can seldom be anticipated from systems of signals that are typically received by decision-makers in the world today. Moreover, the signals that do exist are often not observed, or are wrongly interpreted, or are not part of the incentive structure of societies. This is due to ignorance about the dynamic effects of changes in ecosystem variables (for example, thresholds, buffering capacity, and loss of resilience) and to the presence of institutional impediments, such as lack of well-defined property rights. The development of appropriate institutions depends, among other things, on understanding ecosystem dynamics and on relying on appropriate indicators of change. Above all, given the fundamental uncertainties about the nature of ecosystem dynamics and the dramatic consequences we would face if we were to guess wrong, it is necessary that we act in a precautionary way so as to maintain the diversity and resilience of ecosystems.

Economic growth is not a panacea for environmental quality; indeed, it is not even the main issue. What matters is the content of growth – the composition of

[6] Adaptive management views regional development policy and management as "experiments", where interventions at several scales are made to achieve understanding, to produce a social or economic product, and to identify options. See, for example, Holling (1978), Walters (1986), Lee (1993).

inputs (including environmental resources) and outputs (including waste products). This content is determined by, among other things, the economic institutions within which human activities are conducted. These institutions need to be designed so that they provide the right incentives for protecting the resilience of ecological systems. Such measures will not only promote greater efficiency in the allocation of environmental resources at all income levels, but they would also assure a sustainable scale of economic activity within the ecological life-support system. Protecting the capacity of ecological systems to sustain welfare is of as much importance to poor countries as it is to those that are rich.

References

Anderson, K., & Blackhurst, R. (Eds.) (1992). *The greening of world trade issues.* London: Harvester Wheatsheaf.

Beckerman W. (1992). Economic growth and the environment: Whose growth? Whose environment? *World Development, 20*, 481–496.

Dasgupta, P., & Mäler, K.-G. (1991). The environment and emergent development issues. Proceedings of the Annual Bank Conference on Development Economics, 1990. Supplement to the World Bank Economic Review 1991. Washington DC: The World Bank.

Grossman, G. M., & Krueger, A. B. (1993). In Garber, P. (Ed.), *The U.S. Mexico free trade agreement* (pp. 167–177). Cambridge, MA: MIT Press.

Holling, C. S. (1973). Resilience and stability of ecological systems. *Annual Review of Ecology and Systematics, 4*, 1–23.

Holling, C. S. (Ed.) (1978). *Adaptive environmental assessment and management.* London: Wiley.

Jansson, A. M., Hammer, M., Folke, C., & Costanza, R., (Eds.) (1994). *Investing in natural capital: The ecological economics approach to sustainability.* Washington, DC: Island Press.

Lee, K. (1993). *Compass and the gyroscope.* Washington DC: Island Press.

Lopez, R. (1994). The environment as a factor of production: The effects of economic growth and liberalization. *Journal of Environmental Economics and Management, 27*, 163–184.

Moomaw, W., & Tullis, M. (1994). In Socolow, R., Andrews, C., Berkhout, W., & Thomas, V. (Eds.), *Industrial ecology and global change* (pp. 157–172). Cambridge, UK: Cambridge University Press.

Pimm, S. L. (1984). The complexity and stability of ecosystems. *Nature*, 307, 321–326.

Selden, T. M., & Song, D. (1994). Environmental quality and development: Is there a Kuznets curve for air pollution emissions? *Journal of Environmental Economics and Management*, 27, 147–162.

Shafik, N., & Bandyopadhay, S. (1992). *Economic growth and environmental quality: Time series and cross country evidence (background paper for the World Development Report).* Washington, DC: World Bank.

Vitousek, P., Ehrlich, P. R., Ehrlich, A. H., & Matson, P. (1986). Human appropriation of the product of photosynthesis. *BioScience, 36*, 368–373.

Walters, C. J. (1986). *Adaptive management of renewable resources.* New York: Macmillan.

Chapter 11
Askö 1994: Commentary by Elinor Ostrom

Elinor Ostrom

Attending an Askö meeting is truly a memorable experience. The combination of being in a deeply beautiful setting, that one reaches after a great boat ride, and engaging in a vigorous exchange with outstanding scholars, who are at the forefront of ecology and economics, stimulates the minds of those who attend. The written reports have had an immense impact on the new field of ecological economics. The participants at the Second Askö meeting held in the fall of 1994 set an excellent example. They provided a useful synthesis of the current thinking at the overlap of the borders of economics and ecology as well as stimulating new research questions for the next generation of ecological economists.

Until the early 1990s, many social scientists did not take environmental problems very seriously. As Arrow and his colleagues reflect, the environment was viewed as a "tangential concern and the presumption is often made that economic growth and economic liberalization (including the liberalization of international trade) are, in some sense, good for the environment" (p. 520). Seven years before this meeting, the Brundtland Commission had issued a dramatic call for a major change in the way scholars and policy makers viewed "our common future" (WCED 1987). While not overtly responding to the Brundtland report, the focus of this paper by Arrow et al. did precisely examine the "relation between economic growth and environmental quality, and the carrying capacity and resilience of the environment" (ibid.).

The authors strongly critiqued the presumption held by many scholars that an inverted U-shaped relationship – sometimes called the Environmental Kuznets Curve – existed between the wealth of a country and its environmental quality. While in the early years of economic growth, scholars agreed that environmental degradation would occur, there was hope that as residents gained more wealth, their evaluation of the importance of environmental resources would grow stronger. In this view, economic growth would eventually stimulate a major concern about and action for the environment. Arrow and colleagues argue that while this relationship

E. Ostrom
Workshop in Political Theory and Policy Analysis, Indiana University, USA
e-mail: stodd@indiana.edu

T. Söderqvist et al. (eds.), *Bringing Ecologists and Economists Together*,
DOI 10.1007/978-90-481-9476-6_11, © Springer Science+Business Media B.V. 2011

could be shown to exist in relationship to specific pollutants, they stress that this overconfidence in changing demand patterns did not pertain to the environmental resource base itself. They rightly take on the presumption that the shift in preferences with growing wealth would automatically translate into better environmental policies. They raise the perverse problem that the environmental improvement in one country sometimes leads to deterioration in another country that accepts the environmental trash of the first country.

Arrow and colleagues stressed that the ecological resource base for all economic growth is finite. Institutions well matched to the diverse scales and interrelationships among ecological systems can increase the sustainability of the base over time. Without substantial institutional innovation, however, major ecological degradation is likely to continue, if not accelerate, even as economic growth continues. Thus, one of the most important messages of the article is the need for designing better-crafted institutions at multiple scales. The other critical lesson is that economic growth is not a *panacea* – nor do any *panaceas* exist for enhancing the resilience of the complex multi-scale ecologies so important for our common future. Participants in Askö meetings through the years have now carried this work forward in a variety of research efforts, books, and articles. Readers of this current volume may find the recent work of Fikret Berkes, Steve Carpenter, Marco Janssen, Charles Perrings, and myself in a special feature of the *Proceedings of the National Academy of Sciences* on "Beyond Panaceas" of considerable interest (see references below). These scholars, who have been inspired by the Askö experience, are urging all scientists to use a deeper diagnostic approach when proposing resource policies. We need to stop doing more harm than good due to the casual use of accepted "cure-alls."

References

Berkes, F. (2007). Community-based conservation in a globalized world. *Proceedings of the National Academy of Sciences, 104*(39), 15188–15193.

Brock, W. A., & Carpenter, S. (2007). Panaceas and diversification of environmental policy. *Proceedings of the National Academy of Sciences, 104*(39), 15206–15211.

Ostrom, E. (2007). A diagnostic approach for going beyond panaceas. *Proceedings of the National Academy of Sciences, 104*(39), 15181–15187.

Ostrom, E., Janssen, M. A., & Anderies, J. M. (2007). Going beyond panaceas. *Proceedings of the National Academy of Sciences, 104*(39), 15176–15178.

Perrings, C. (2007). Future challenges. *Proceedings of the National Academy of Sciences, 104*(39), 15179–15180.

World Commission on Environment and Development (WCED). (1987). *Our common future.* New York: Oxford University Press.

Chapter 12
Askö 1994: Commentary by V. Kerry Smith

V. Kerry Smith

One of the most durable questions about how the economy is linked to the natural environment asks whether the scale of economic activity is limited by the availability of natural resources. Generations of economists have struggled with this question.[1] Does a world without oil, for example, foreshadow a future of decline or merely another transition in the kaleidoscopic change facilitated by technological innovation?[2]

The character of these discussions changed with the emergence of environmental economics and the research of Allen Kneese and John Krutilla at Resources for the Future in the 1960s and 1970s. They forced economists to acknowledge that physical laws govern production and consumption activities.[3] Residuals are inevitable consequences of both production and consumption. We are challenged to dispose of them in ways that don't impinge on the capacity of natural systems to provide disposal and amenity services in the future. While much has improved in developed economies over the past 35 years, conventional disposal practices still lead to significant pollution and externalities. Moreover, some of today's decisions

---

[1] See Barnett and Morse (1963) for a classic study that reviews this early literature and addresses the question by measuring the trends in the labor and capital required to obtain a unit of extractive output. Their analysis of real unit cost has been widely cited as "testing" scarcity. For an early follow-up revisiting the issue, see Smith (1979). A more recent re-appraisal is given in Simpson et al. (2005).

[2] This characterization is taken from Barnett and Morse's summary of why in their view resources had not limited economic activity. They concluded, "...the resource spectrum undergoes kaleidoscopic change through time. Continual enlargement of the scope of substitutability – the result of man's technological ingenuity and organizational wisdom – offers those who are nimble a multitude of opportunities to escape" (p. 244).

[3] Summaries of the Kneese and Krutilla research programs that are relevant to these arguments can be found in Kneese and Bower (1979) and Krutilla and Fisher (1975).

V.K. Smith
Department of Economics, W.P. Carey School of Business,
Arizona State University, USA
e-mail: kerry.smith@asu.edu

T. Söderqvist et al. (eds.), *Bringing Ecologists and Economists Together*,
DOI 10.1007/978-90-481-9476-6_12,

about how to use our environmental resources have irreversible consequences. With incomplete information, careful evaluation of these decisions is especially important.[4]

Kneese and Krutilla's research offered the first step in transforming what is often characterized as a fundamental description of economic activity – *the circular flow model*. One version of this diagram is given in Fig. 12.1. It depicts the parallel real and monetary flows that characterize the key components of economic activity. In this model, the important flows of goods and services from firms to households arise from trades in markets. Output markets trade final goods and services. The payments for them define household expenditures and revenues for firms. Trades on factor markets give rise to prices that influence firms' costs and determine households' incomes. Many conventional economists consider this setting as providing all the information that is important about people's behavior. What is remarkable about this picture, from the perspective of the natural scientists and increasingly environmental economists, is that it is a story that omits the very systems that sustain human existence!

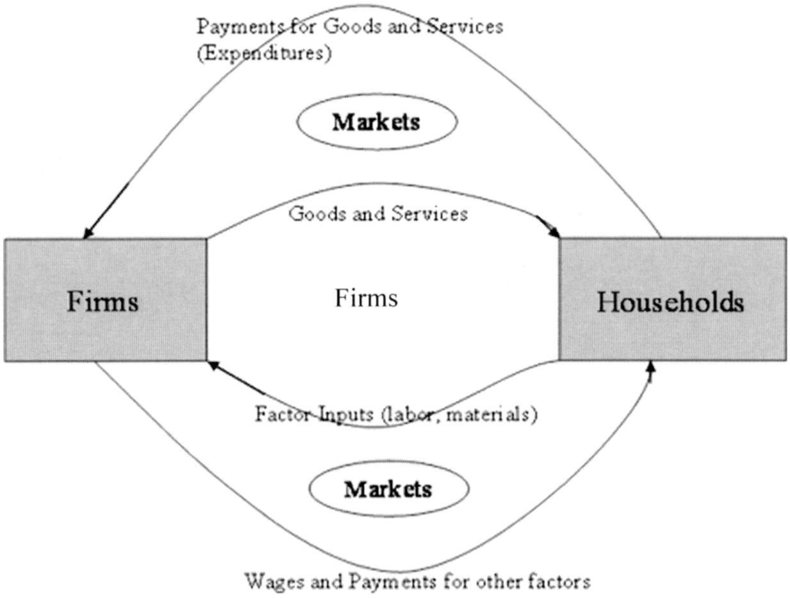

Fig. 12.1 The circular flow

[4] Arrow and Fisher (1974) first demonstrated the importance of irreversibility and uncertainty when there is the prospect for learning new information over time. Their analysis was motivated by Krutilla's (1967) early arguments suggesting that irreversible changes should alter the criteria used in public investment discussions.

Arrow et al. challenge this conception of the fundamentals describing economic activity. To properly account for what is important about economic processes, they must be integrated within the environmental systems that underlie all human activity. Flows into and out of the conventional circular flow's depiction of how "nature" interacts with the economy don't capture the essence of Arrow et al.'s argument. Their model is *not* one in which two systems operate in parallel, making periodic exchanges that are incidental to economic and ecological activities. Rather, the Arrow et al. conception envisions a single, integrated system where the non-market feedback effects of pollution alter demand and supplies for market goods which, in turn, change the sources of those feedbacks. This is nothing short of joint determination. It is equally relevant in describing static general equilibrium models or in characterizing an evolving system in dynamic growth models.

How does this perspective change the way we answer the question of limits to the scale of economic activity? The Arrow et al. answer departs from much of the literature attempting to judge resource scarcity. These authors are direct in their response. We won't be able to answer the question of whether environmental and resources services are growing dangerously scarce in time to make a difference! Moreover,

> "...given the fundamental uncertainties about the nature of ecosystem dynamics and the dramatic consequences we would face if we were to guess wrong, it is necessary that we act in a precautionary way so as to maintain the diversity and resilience of ecosystems. (p. 521)"

Has their warning and the call for institutional reform to signal the impacts we are having on ecosystems been heeded? Unfortunately, in my view, the answer is no. There are more advocates for their position. Ten plus years later, more academics are sounding alarms. Concerns over consequences of global warming, for example, are penetrating the media with increasing frequency, but policy reform is nonexistent, at least in the United States. Dramatic growth in China and India largely undermines efforts in other developed economies to address the global consequences of growth. We do not have institutions that require private users of ecological systems' services to take into account the social consequences of their actions.

Surely this observation seems to convey a gloomy outlook. Perhaps it should not be interpreted in this way. Change in human activities can be rapid if we are able to provide sustained incentives for people to change their behavior. Consider some things we take for granted in developed economies. Less than 100 years ago, the electric lighting technology reached the efficiency of most of today's light bulbs. As Nordhaus (1997) documents, that in under 200 years (between 1800 and 1992) lighting efficiency increased by a factor of 900.[5] We did not have a coordinated system of keeping time until 1911 (Prerau 2005). Now the world is synchronized at a high level of temporal resolution.

[5] Table 1.3 in Nordhaus's paper compares the efficiency of different lighting technologies and documents the approximate date of their introduction.

Continuous incentives assure a sustained commitment. That is, why it is especially important to focus attention on papers such as this one. The policy process needs to take the next steps to assure the economic systems that emerge during the next century transform the nineteenth century model of economic activity. We must conceptualize economic activity in a new way. It cannot be represented in a "flat" picture of a single circular flow. The economic world may be flat, as Thomas Friedman (2005) suggests, but the *economic ecology* of human activities requires we describe these interactions in "spherical terms." They are part of a truly "global flow" with markets one part of the large set of systems that determine how economic growth and the environmental change will evolve over time.

References

Arrow, K. J., & Fisher, A. C. (1974). Environmental preservation, uncertainty and irreversibility. *Quarterly Journal of Economics, 88*, 312–319.

Barnett, H. J., & Morse, C. (1963). *Scarcity and growth*. Baltimore, MD: Johns Hopkins University for Resources for the Future.

Friedman, T. F. (2005). *The world is flat: A brief history of the twenty-first century*. New York: Farrar, Straus, and Giroux.

Kneese, A. V., & Bower, B. T. (1979). *Environmental quality and residuals management*. Baltimore, MD: Johns Hopkins University for Resources for the Future.

Krutilla, J. V. (1967). Conservation reconsidered. *American Economic Review, 57*, 777–786.

Krutilla, J. V., & Fisher, A. C. (1975). *The economics of natural environments: Studies in the valuation of commodity and amenity resources*. Baltimore, MD: Johns Hopkins University for Resources for the Future.

Nordhaus, W. D. (1997). Do real output and real wage resources capture reality? The history of lighting suggests not. In T. F. Breshnahan & R. J. Gordon (Eds.), *The economics of new goods*. Chicago: The University of Chicago Press for the National Bureau of Economic Research.

Prerau, D. (2005). *Seize the daylight: The curious and continuous story of daylight savings time*. New York: Thunder's Mouth Press.

Simpson, R. D., Toman, M. A., & Ayres, R. U. (Eds.). (2005). *Scarcity and growth revisited: Natural resources and the environment in the new millennium*. Washington, DC: Resources for the Future.

Smith, V. K. (Ed.). (1979). *Scarcity and growth reconsidered*. Baltimore, MD: Johns Hopkins University for Resources for the Future.

Chapter 13
Askö 1995: Resilience in Natural and Socioeconomic Systems*

Simon A. Levin, Scott Barrett, Sara Aniyar, William Baumol, Christopher Bliss, Bert Bolin, Partha Dasgupta, Paul Ehrlich, Carl Folke, Ing-Marie Gren, C.S. Holling, AnnMari Jansson, Bengt-Owe Jansson, Karl-Göran Mäler, Dan Martin, Charles Perrings, and Eytan Sheshinski

---

*This article was published in *Environment and Development Economics*, Vol. 3, Levin et al., Resilience in natural and socioeconomic systems, 222–235, Copyright Cambridge University Press (1998).

S.A. Levin
Department of Ecology and Evolutionary Biology, Princeton University, Princeton, NJ 08544, USA

S. Barrett
London Business School, Sussex Place, Regent's Park, London, NW1 4SA, UK

S. Aniyar
Facultad de Ciecias Economicas y Sociales, Universidad del Zulia, Maracaibo, Venezuela

W. Baumol
CV Starr Center for Applied Economics, New York University, 269 Mercer Street, New York, NY 10003, USA

C. Bliss
Institute of Economics and Statistics, University of Oxford, Oxford, OX1 3UL, UK

B. Bolin
Department of Meteorology, Stockholm University, 106 91 Stockholm, Sweden

P. Dasgupta
Faculty of Economics, University of Cambridge, Sidgwick Avenue, Cambridge CB3 9DD, UK

P. Ehrlich
Department of Biological Sciences, Stanford University, Stanford, CA 994305, USA

C. Folke
Department of Systems Ecology, Stockholm University, S-106 91 Stockholm, Sweden and Beijer Institute of Ecological Economics, the Royal Swedish Academy of Sciences, Box 50005, SE-10405 Stockholm, Sweden
e-mail: carl.folke@beijer.kva.se

I.-M. Gren
The Beijer Institute of Ecological Economics, the Royal Swedish Academy of Sciences, Box 50005, SE-104 05, Stockholm, Sweden

T. Söderqvist et al. (eds.), *Bringing Ecologists and Economists Together*,
DOI 10.1007/978-90-481-9476-6_13, © Springer Science+Business Media B.V. 2011

13.1 Introduction

We, as a society, find ourselves confronted with a spectrum of potentially catastrophic and irreversible environmental problems, for which conventional approaches will not suffice in providing solutions. These problems are characterized, above all, by their unpredictability. This means that surprise is to be expected, and that sudden qualitative shifts in dynamics present serious problems for management. In general, it is difficult to detect strong signals of change early enough to motivate effective solutions, or even to develop scientific consensus on a time scale rapid enough to allow effective solution. Furthermore, such signals, even when detected, are likely to be displaced in space or sector from the source, so that the motivation for action is small. Conventional market mechanisms thus will be inadequate to address these challenges.

To deal with such problems, one needs a response system that is flexible and adaptive. What principles might underlie the development of such a system? In drawing from experience, it is useful to examine patterns in how systems or institutions respond to stress. What are the success stories and failures, and what can we learn from them? Is there a thread that connects the response of the human immune system to infectious agents, sociological dynamics that lead to the collapses of regimes and forms of government, or ecological phenomena such as responses to the depletion of stratospheric ozone or the desertification of productive savannas? All exhibit the characteristic behaviors of non-linear dynamical systems, interacting across different scales. Hence all exhibit the potential for thresholds, for multiple domains of stability, and for path dependency. Thus, none can be treated by

A. Jansson
Department of Systems Ecology, Stockholm University, S-106 91 Stockholm, Sweden

K.-G. Mäler
Beijer Institute of Ecological Economics, the Royal Swedish Academy of Sciences, Box 50005, SE-104 05 Stockholm, Sweden

B.-O. Jansson
Department of Systems Ecology, Stockholm University, S-106 91 Stockholm, Sweden

C.S. Holling
Department of Zoology, University of Florida, 111 Bartram Hall, PO Box 118525, Gainesville, FL 32611–8525, USA

D. Martin
The John D. and Catherine T. MacArthur Foundation, 140 South Dearborn Street, Chicago, IL 60603, USA

C. Perrings
Environment Department, University of York, Heslington, York, YO1 5DD, UK

E. Sheshinski
Department of Economics, The Hebrew University of Jerusalem, 20 Tura Street, Yemin Moshe, Jerusalem 91905, Israel

traditional markets, or regulatory policies. All require the development of flexibility and the capacity for adaptive response.

In ecological and socioeconomic systems alike, human activities can lead to qualitative shifts in structure and function, evidence that the system concerned has lost resilience: that it is no longer capable of absorbing the stresses and shocks imposed by human activity without undergoing a fundamental change involving loss of function and, often, loss of productivity. Resilience, the ability to experience change and disturbance without catastrophic qualitative change in the basic functional organization, is a measure of the system's integrity (Holling 1973).

Ecological and economic systems are non-linear and adaptive, exhibiting complex and far-from-equilibrium dynamics. Much classical theory in both disciplines, however, relies heavily on locally linear methods to estimate stability and times of return to equilibria, thereby ignoring many phenomena of overriding importance such as the potential for catastrophes and domain shifts.

Most economic analyses of resource management schemes,[1] for example, begin by estimating exploitation costs in relation to the quantity exploited and the existing stock, and then calculating the social benefits from exploitation as a function of the amount exploited and possibly the remaining stock of the resource. From this, one can compute the desired time profile of exploitation as simply the time profile that maximizes the (expected) present value of (future) net benefits, i.e. the difference between costs and benefits. For non-renewable resources, with relatively simple dynamics, this works well; for living resources, however, far greater non-linearities intrude, complicating analysis immensely. In particular, simply quantifying local balances in the neighborhood of a steady state may be profoundly misleading.

Complex non-linear systems, in general, typically are characterized by multiple domains of attraction; hence historical contingency has a large influence on their behaviors. Evolutionary change through natural selection and speciation, for example, is profoundly influenced by historical accident, so that evolutionary theory is far more powerful as a retrospective tool than for purposes of detailed prediction (Jacob 1977; Slobodkin 1964). Similarly, the dynamics of biological communities over ecological time scales are highly non-linear, and the existence of multiple stable states both influences our understanding of current configurations, and raises concerns about possible qualitative shifts in the face of large human-induced or natural perturbations (Holling et al. 1995).

Overgrazing can, for example, lead to desertification; and overexploitation, to the collapse of fisheries and other resource systems. Another example is provided by changing land-use patterns and consequent habitat fragmentation, leading to dramatic species losses with serious consequences for the maintenance of ecological services. Similarly, increased human population size, consumption of resources, technological innovation and mobility can engender pest outbreaks, including the spread of devastating human diseases, linking ecological and health factors intimately.

[1] There are of course exceptions, for example in fishery economics where even chaotic behavior of fish stocks have been taken into account.

In socioeconomic systems, war and civil unrest, representing the break-down of a balance of peace and security, present striking examples of qualitative shifts in which initially small disturbances may become magnified through non-linear feedbacks. The Great Depression of 1929 provides another example of a sudden collapse from one 'basin of attraction' into another; furthermore, even more puzzling than the initial collapse of the markets was why the economy remained trapped at low levels until a new perturbation, the Second World War, stimulated markets and initiated recovery. Ecological and socioeconomic systems that are persistent and stable in the face of historical patterns of disturbance may lack the resilience to withstand such novel disturbances, especially those of large magnitude, and the potential consequences become matters of deep concern to the conservation of our ecological and socioeconomic environment.

Resilience thus makes no distinctions, preserving ecologically or socially undesirable situations as well as desirable ones. It helps maintain our environments, sometimes leading to a too-cavalier attitude about the robustness of our life support systems; it similarly translates into resistance to change when such change is mandated. The existence of multiple stable regions of attractions further implies that such regions will be separated by less 'attractive' dynamical regimes, both in a dynamical sense and often in the sense of public policy. Consider, for example, the familiar but illustrative metaphor of the system state as represented by a ball that becomes pushed from one valley to another, necessarily passing along the way over a ridge that is higher than either valley. The resilience of systems is governed by both the height of such ridges and the distances between them. The transition between zones of attraction is often a painful one; a good example is the dislocation caused by efforts to shift economic systems, such as in the former Soviet Union, from one zone to another.[2]

Not all resilient phenomena are desirable. For example, discriminatory class systems have proved resilient. Similarly, racism has proved stubbornly resistant to policies aimed at wrecking its foundations. Though some social systems are too fragile, others are too hardy. A similar phenomenon may be observed at the level of the household. High rates of fertility in the third world, for example, are caused in part by the absence of effective capital markets, which make it difficult or impossible for households to obtain social security except by having more children. In turn, increases in population density facilitate the transmission of disease; and as families become more vulnerable to disease they may respond by having yet more children, so as to be sure of having more surviving members. This can be a vicious feedback, and one requiring effective public action, not just in terms of immunization and disease control but also in providing for the security of citizens in their old age (Dasgupta 1995).

There are, however, many new threats to the resilience of human institutions and the ecological systems on which they depend. Examples include global climate change, the loss of biodiversity, and the emergence of new infectious diseases. If the climate, certain natural environments, and the human immune system were

[2] This example was provided by K. Arrow, personal communication.

resilient to all perturbations, changes in the atmospheric concentration of greenhouse gases or biodiversity loss or the incidence of disease would have small consequences for human well-being, except for the often-important time delays that would be involved in recovery. But since these systems are non-linear, the sum total of our actions may cause them to lurch to quite different, and potentially unwelcome, stability domains. Thus, small changes in global mean temperature could lead to a shift in the Gulf Stream; extinction of certain keystone species may magnify into losses of ecosystem function (Levin 1997); and the increased and inappropriate use of antibiotics and the presence of more immune-compromised people because of malnutrition, exposure to harmful pollutants, and the HIV virus may erode the capacity of human beings to resist disease.

Similarly, in economic systems, small or medium sized disturbances may be beneficial for the growth of productivity. Schumpeter (1912) created a theory of economic development based on this observation. His famous term *creative destruction* was coined to describe the window for novelty and creation that was generated by the failures of existing plants with their old technologies. Thus, according to Schumpeter, it is all these disturbances that in the long run, through the creative entrepreneurs, create economic growth and an ability to survive major changes such as economic depressions. Similarly, the concept of *X*-efficiency (Leibenstein 1966) introduced by Leibenstein can be interpreted in a similar way. Companies that live in a very stable, protected environment will not have incentives to develop the flexibility and competitiveness that would be needed if a major shock were to hit the market. Companies that always have to fight for survival develop resilience much more fully, partly by the necessity to increase productivity. It is striking the degree to which such theories of economic resilience parallel concepts in evolutionary biology, in which similar tradeoffs have been deeply explored.

13.2 The Nature of Resilience

One of the central features of non-linear systems, and one that confounds management, is that small perturbations can become magnified and lead to qualitatively unexpected behaviors at more macroscopic levels; this becomes increasingly true as system complexity increases. In the case of forest fires, for example, the gradual build-up of fuel during periods without fires can provide the ingredients that can lead to major conflagrations. Classical management regimes endeavored to suppress fires entirely, apparently preserving the status quo in terms of the most evident variables. Yet total fire suppression increases both the stock of timber, and thus the stock of combustible litter, creating accidents waiting to happen. Through simplistic management regimes, robustness and resilience are lost, and the predisposition to catastrophic fires is increased. Such fires would spread farther and burn longer and at higher temperature than otherwise, leading not just to the destruction of trees and seeds, but also to soil erosion and a deterioration in the capability of the system to recover. It was for example the myopic success of earlier fire control that made the recent fires in Yellowstone National Park so devastating. Enlightened forest management now recognizes the value of letting small fires burn in order to maintain the resilience of systems. Similarly, profligate use of antibiotics to treat individuals provides short-term gains, but increases overall susceptibility of populations to catastrophic disease outbreaks because of selection for more resistant pathogens (Lederberg et al. 1992; Daily and Ehrlich 1996).

Domain shifts may be virtually permanent, as in desertification or fisheries (Steele 1996), or periodic, as for example in the recurrent outbreaks of forest pests or diseases such as influenza. In evolutionary theory, attention to punctuated equilibria recognizes the potential importance of permanent domain shifts on even larger time scales.

In the case of AIDS, one theory (e.g., Nowak et al. 1991; but see Wolinsky et al. 1996) argues that a domain shift occurs (an infected individual proceeds to full-blown AIDS) when the capacity of the immune system to repel new mutants is overcome by the diversification of strains through mutation. An analogous phenomenon perhaps underlay the sudden collapse of communism in Central and Eastern Europe. In those nations, although there had always existed a small fringe of political dissidents, persistent and growing economic difficulties strengthened and diversified the voice of opposition. Importantly, the larger is the opposition to any social system, the less likely it is that any one dissident will be singled out. So as the outer fringe of the population began to express their opposition to communism, this in turn gave strength to their 'neighbors' to become vocal, which in turn gave their 'neighbors' the strength to become vocal, ... and so forth; until a sufficiently large proportion of people were 'out on the streets', as it were, forcing a change in political regime. (This is an example of 'tipping'. For a readable account, see Schelling 1978.) In democracies, political institutions have the broad consent of the people they represent, and elected officials are periodically held to account at the polls. This makes governments responsive to stresses and shock and hence more resilient, which also enhances the potential for gradual rather than catastrophic change.

The important destabilizing consequences of global change will be felt locally; but only after they have been processed through changes at the scale of ecosystems, and landscape groupings of ecosystems, where a history of inappropriate resource development and exploitation has already eroded resilience. This, combined with the unavoidable delays in relating effects to causes, will lead to an even greater loss in local resilience. At the same time, because global environmental change is caused by the actions of a large number of parties, most of which are far removed from any one locality, local institutions are rendered even less capable of rebuilding resilience.

What this suggests is that policy should be concerned with more than the immediate consequences of incremental actions. It should recognize the potential for an accumulation of small actions, each on their own perhaps quite harmless, to destabilize important natural and social systems. The difficulty is that, while we can predict with reasonable confidence the immediate consequences of an incremental action, we cannot predict the consequences of an entire sequence of actions without understanding the systems potentially being affected by them.

Every natural system is subject to regular disturbances; those that have survived indeed must have built up some degree of resilience. Ironically, however, the mechanisms that provide short-term resilience may also impose a rigidity of structure that erodes the capacity to respond to disturbances over longer time scales, leading to punctuated changes through an increasing loss of buffering capacity. Resilience comes from flexibility, and the ability to change adaptively. From an evolutionary point of view, we can take lessons from the prototypical adaptive management system, the human immune system, which has evolved to deal with a wide variety of unanticipated challenges by effectively training itself to respond to threats. In somewhat the same spirit, biospherists have become enamored with GAIA, the notion that the biosphere has its own self-regulatory mechanisms that will protect it against new threats. There is an attractiveness in this notion, since as already argued the very existence of the biosphere as we know it means that it must be to some extent buffered against the environmental challenge that might have destroyed lesser systems. Yet the theory in its boldest form is flawed in that the biosphere is a unique realization of its evolutionary history, not the result of selection among multiple candidates for the capability to respond to stress. It is self-organized rather than selected, the result of the evolution of its components. Much of its buffering capacity in the face of perturbations is encoded in part in its biodiversity, but we have very little understanding of what the quantitative relationships are between loss of species and loss of ecosystem function (Levin 1997).

Any complex system will exhibit hierarchical organization, in which dynamics at one level, including especially stability and resilience, emerge from phenomena taking place at lower levels of organization. Social systems are no exception. At the level of the household, adaptability has come from the way in which members of the household integrate with the wider community. A household may insure itself against the risk of an illness or accident befalling any of its members by means of the extended-family system. Engaging in transactions outside the family can enhance resilience even further. Most obviously, where risks are household specific, pooling risks across households can reduce substantially the risk any one household faces. Trade between regions can also be welfare-improving, not only because it can allow regions to exploit their comparative advantage and thus increase incomes, but also because it allows for further risk spreading. For example, cattle in India are sold in regional rather than village markets, allowing households to smooth their consumption against village-specific risks such as drought or floods or disease (Dasgupta 1993). It also can reduce resilience, i.e., by reducing food security in some nations. The potential for risk spreading through trade is also dependent on ecosystem resilience to secure the capacity of the environment to generate a flow of essential natural resources and ecosystem services. This capacity is rarely reflected in the price signals of international trade.

Government also can supply goods and services that individuals would or could not provide themselves in sufficient quantities. Immunization is a good example. If a child is immunized against measles, the population as a whole become more protected. Parents, however, have little or no incentive to take this benefit into account when immunizing their own child, with the result that all families may be better off with state-sponsored mandatory immunization.

What makes for a resilient system in general? Ecological systems typically can be thought of in terms of a hierarchy of responses (Slobodkin 1964). Small perturbations can be accommodated through the behavioral or physiological responses of individuals; stronger stresses lead to reduced reproduction or death and changes at the population level. Ultimately, over generations, natural selection and other forces will change the genetics of populations. The hierarchy of responses may also be seen in organizational responses to perturbation. When effective, they buffer the system against qualitative changes; but all negative feedback systems, most notably the human immune system, may fail. The lessons are obvious for a broad class of environmental and socioeconomic systems. The critical features are a hierarchy of feedback mechanisms; the maintenance of diversity; options for selection to act upon; and the coupling of stimulus and response in terms of space, time, and organizational scales. These are necessary conditions, but do not guarantee success.

In economic systems, resilience depends on comparable mechanisms: effective feedback mechanisms; the coupling of stimulus and response; and a diversity of resources. Three ingredients are important to the first two: competition, effective government, and other effective institutions (North 1990). The third depends on the social investment portfolio and on the treatment of risk and uncertainty. It depends on the effectiveness of mechanisms both to spread and to contain risks, and to maintain options to respond constructively and creatively to shocks. An essential element in all three is trust.

13.3 The Challenge for Management

In managing such complex systems, it is important to realize the potential and limitations for resilience, and to design complementary mechanisms that increase resilience in systems such as the biosphere that are essential to human survival and welfare. To some extent, competition can help build resilience in such systems by means of the feedback mechanism provided by the market. This coordinates the activities of millions of people such that, after an economy is subjected to a parameter change (and assuming certain conditions hold; see below), resources are reallocated until all possibilities for mutual gain have been exhausted. In the event of climate change, for example, market forces would compel farmers to alter their use of inputs and provide incentives for firms to develop new seed varieties, such that the extent of damage from climate change is reduced.

This view that market forces are sufficient to maintain systems as we need them is, however, as naive as assuming that the biosphere's self-regulating potential assures its robustness. Market forces will not function well if unsupported by effective institutions. The market allocates resources guided by prices. If the institutional framework for markets does not account for the increasingly scarce capacity of the environmental resource base to provide support to the socioeconomy, the allocation by markets will be suboptimal, and may even enhance ecosystem deterioration. A self-organized system will achieve its own asymptotic behavior, not one that need correspond to optimization in any desirable way. Despite theories to the contrary, ecosystems achieve patterns that emerge from selection at lower levels of organization, not that optimize aspects of ecosystem function. It has, furthermore, often been observed by evolutionary theorists that optimality is a state, while adaptation is a process (Lewontin 1977) and that one need not lead to the other. In the elegant metaphor of Jacob (1977), evolution is the result of tinkering, not of grand design. If we hope to use market forces to achieve broader goals, we must change the institutional framework, the norms and rules to steer the process. We must, in short, tighten feedback loops. Unless firms are given the property rights to their intellectual achievements (such as the new seed varieties noted above) in the form of patent protection, for example, the market will fail to exploit all opportunities for mutual gain. This is not to recommend private property as a cure-all. For example, a substantial portion of the wealth of poor people in developing countries is in shared or common property resources (Ostrom 1990; Hanna et al. 1996). If they are privatized the people who depend on them may suffer a loss in wealth, and this loss may in turn deprive these people of the wherewithal to participate meaningfully in economic life; they become destitutes (Dasgupta 1993). The result may not only be inequitable but also slow the rate and change the content of economic growth and impair resilience. By 'effective' institutions we mean instead a wide variety of institutions that sustain, by diverse mechanisms, multiple outcomes.

Where competitive markets do not function well the problem typically lies in inappropriate prices or missing markets, and these in turn are often the consequence of inappropriate property rights and badly designed institutions. Whenever anthropogenic emissions cause greenhouse gas concentrations in the atmosphere to increase, for example, the climate can be modified in ways that are deleterious to the human exercise. But the individuals who emit these gases do not presently pay for this damage at a level correlated with their influence; to them, the emission of greenhouse gases is free (prices are 'wrong', the importance of the environmental resource base is not captured in the prices).

The prices are wrong because there are no institutions providing the preconditions for market atmospheric concentrations (the market is 'missing'). Correcting the 'market failure' is not so easy in this case. The atmosphere cannot be nationalized, and countries collectively have not thus far succeeded in devising an international regime for regulating the emission of greenhouse gases.

Plainly, then, the reach of markets and governments does not always extend as far as is necessary to build the desired resilience in social systems. This is why trust, too, is needed. Evolutionary biologists understand that the simplest form of cooperation is reciprocal altruism, in which individuals achieve a fragile trust with one another. The same applies to the mechanisms needed for environmental protection. Trust is a mechanism for widening the basic unit of self-interest, and thus tightening diffuse feedback loops. One party will not contract with another unless it can trust this party to fulfill its obligations. Though government institutions can enforce contracts, contracts cannot account for every possible contingency, and the contracting parties must then trust each other if they are to sustain a mutually desirable outcome in the wake of some shock. Trust in these instances can work through the reciprocal nature of the relationship. Your neighbor may help you today, in the belief that if he does so you will assist him tomorrow if he is in need, and so on. Reciprocal altruism is an important stabilizing force, and its evolution and maintenance are enhanced by the local nature of interactions.

Similar risk-sharing rules are needed to manage the global commons. There is no world government, and so trust between nations is particularly important to making cooperative regimes for managing the global commons resilient. It is remarkable that, despite the absence of a central authority, states comply with their treaty obligations in the vast majority of cases (Chayes and Chayes 1995). What seems to sustain cooperation is that failure to do so can have system-wide consequences. This restrains states from behaving contrary to established norms, and this in turn builds a degree of trust.

But trust is not likely to be enough to protect the global environment. Though customary law obligates nations to comply with the international agreements they endorse, it does not obligate nations to endorse any particular agreement. As a consequence, treaties that would be mutually beneficial may fail to attract any support. Alternatively, they may attract universal support but at the cost of requiring that the parties to it do little more than they would have done in the absence of the agreement, or they may require greater sacrifices but accordingly succeed in attracting only a small number of participants. Of course, international cooperation is not always hard to sustain. But it can be expected to be so for problems like global climate change where the benefits of, say, a stabilization of atmospheric concentrations of greenhouse gases would be shared by all nations and where the costs of effecting such a stabilization would be high for each nation. There will then be a temptation to free-ride.

Admittedly, the failure of cooperation need not be disastrous, because some feedback loops are closed locally. Countries will almost always have an incentive to take some actions unilaterally. In the case of global climate change, for example, a number of countries have introduced a carbon tax. This has created incentives not only for individuals and firms to reduce their carbon emissions but also for industry to innovate – to fund research into the alternatives to fossil fuels and to develop the technologies that will reduce the costs of abatement over time. Subsidies to R&D would of course have a similar effect. Governments can also reduce the local consequences of global climate change by providing local public goods like barriers to rising sea levels. All of these unilateral responses can be important. But they can also fall far short of what all nations would prefer was actually done about the problem.

An important exception is the Montreal Protocol, the international treaty governing the use of ozone-depleting substances like CFCs. The agreement is remarkable, not least because it is phasing out these substances in a relatively short time, and with the participation of nearly every country in the world. The agreement has also adapted to changing circumstances. It has been renegotiated on several occasions, prompted by changes in either the science or economics of ozone depletion. And it has been designed to be stable with respect to possible free-riders. It appears that, because of the trade sanctions allowed by the treaty, defection by even a number of parties would not trigger an unraveling of the agreement. Having the assurance that the treaty will not dissolve reinforces the incentives which countries have to measure up to the obligations of the treaty. This in turn creates a situation in which renegotiation can take place, as changing circumstances warrant.

Unfortunately, the Montreal Protocol is a rarity in this regard. International agreements have not proved as successful in other areas, like the management of international fisheries and the global climate (Barrett 1990). Similarly, while the eradication of smallpox stands out as one of the greatest achievements of this century, we have made comparatively little progress in controlling malaria, cholera, tuberculosis, polio, and measles, especially in developing countries. There is no reason to believe that the appropriate resilience will arise 'spontaneously'.

In general, resilient systems typically have a hierarchy of feedback mechanisms, and the maintenance of heterogeneity and diversity for selection to act upon. Heterogeneity in space or in sector has a number of implications. It implies, first of all, that there is the requisite diversity, and the potential for change. It also means that responses initially will be localized both in cause and effect, so that concern must exist for providing the motivation for local responses in a global commons in ways that benefit the common good, and for the spread of socially desirable innovations.

The implications of any action need to be examined carefully before we become committed to its consequences. Particular attention needs to be devoted to indirect effects. It was the indirect effect of fire control making available a larger stock of biomass for burning that rendered that policy harmful. In the case of global climate change, much discussion has centered on mean global temperature rise and not the effects that may be triggered by this warming, such as a sudden change in the course of the Gulf Stream. Linkages between resilience in natural and social systems also demands more research. As an example, local control over the Mudialy wetlands near Calcutta appears to have restored the productivity of the fishery and produced a harvest of wood in addition (Pye-Smith and Feyerabend 1994). A diversity of such ecologically adapted management practices and linked social mechanisms behind these practices for building resilience are found in both local and regional systems of traditional and contemporary societies (Berkes and Folke 1998).

Research needs to address whether or not our vital systems have sufficient capacity for resilience. Building resilience into our natural systems means maintaining crucial ecosystem elements, such as the soils. The scale of the human enterprise is important, but it is not the level of economic activity itself that determines resilience. Equally important is the nature of this activity: it matters whether profligate consumption of fossil fuels continues or innovative steps are taken to develop fossil fuel substitutes. At the institutional level, resilience requires a competitive economy, effective government, and trust. The global environment is especially in need of resilience, and yet it is precisely at this level that government institutions are least effective and trust most delicate.

Critical scientific and economic conclusions about resilience need to be made available to decision-makers and the public at large in a way that aids understanding.

The institutional apparatus that exists now is not always well suited to this. Lobbying groups convey information, but none has an incentive to do so in a balanced way, and it is not clear that having different lobbyists arguing opposite positions contributes to understanding; often this arrangement only polarizes opinion. This is the value in conservative organizations like the Intergovernmental Panel on Climate Change. The IPCC is made up mostly of university-based natural and social scientists, nominated by their governments. Its aim has been to clarify the issues and find a consensus among the global scientific community; and it is noteworthy that the IPCC has sought to involve as many countries as possible in this process. Having a common understanding of a shared problem is a vital ingredient of trust; which is itself a basic building block of resilient social systems, and these will help humanity to interact with natural systems without degrading them.

Most importantly, we must appreciate better the degree to which ecological and socioeconomic systems are linked, across a range of scales of space, time, and complexity. They must be viewed as one single system. We can only ultimately slow the rate of change and the appearance of surprises by slowing population growth, energy use, and the rate at which we are changing our environment. We must rise above parochial interests and tribalism, developing a new order and new mechanisms for providing local, regional, and global incentives that will allow our life support systems to persist for future generations.

References

Barrett, S. (1990). The problem of global environmental protection. *Oxford Review of Economic Policy, 6*, 68–79.

Berkes, F., & Folke, C., (Eds.) (1998). *Linking social and ecological systems: Management practices and social mechanisms for building resilience*. Cambridge: Cambridge University Press.

Chayes, A., & Chayes, A. H. (1995). *The new sovereignty*. Cambridge, MA: Harvard University Press.

Daily, G. C., & Ehrlich, P. R. (1996). Impacts of development and global change on the epidemiological environment. *Environment and Development Economics, 1*, 311–346.

Dasgupta, P. (1993). *An inquiry into well-being and destitution*. Oxford: Clarendon.

Dasgupta, P. (1995). The population problem: Theory and evidence. *Journal of Economic Literature, 33*(4), 1879–1902.

Hanna, S. S., Folke, C., & Mäler, K.-G. (Eds.). (1996). *Rights to nature: Ecological, economic, cultural, and political principles of institutions for the environment*. Washington, DC: Island Press.

Holling, C. S. (1973). Resilience and stability of ecological systems. *Annual Review of Ecology and Systematics, 4*, 1–24.

Holling, C. S., Schindler, D. W.,Walker, B. W., & Roughgarden, J. (1995). Biodiversity in the functioning of ecosystems: an ecological synthesis. In C. Perrings, K.-G. Maler, C. Folke, C. S. Holling, & B.-O. Jansson (Eds.), *Biodiversity loss: Economic and ecological issues. Cambridge*: Cambridge University Press.

Jacob, F. (1977). Evolution and tinkering. *Science, 196*, 1161–1166.

Lederberg, J., Shope, R. E., & Okads, S. C., Jr. (Eds.). (1992). *Emerging infections: Microbial threats to health in the United States*. Washington, DC: National Academy Press.

Leibenstein, H. (1966). Allocative efficiency vs. *X*-efficiency. *American Economic Review, 56*, 392–415.

Levin, S. A. (1997). Biodiversity: Interfacing populations and ecosystems. In T. Abe, S. A. Levin, & M. Higashi (Eds.), *Biodiversity: An ecological perspective* (pp. 277–288). New York: Springer.
Lewontin, R. C. (1977). Adaptation. *Enciclopedia Einaudi Turin, 1*, 198–214.
North, D. C. (Ed.). (1990). *Institutions, institutional change and economic performance*. Cambridge: Cambridge University Press.
Nowak, M. A., Anderson, R. M., McLean, A. R., Wolfs, T., Goudsmit, J., & May, R. M. (1991). Antigenic diversity thresholds and the development of AIDS. *Science, 254*, 963–966.
Ostrom, E. (1990). *Governing the commons: The evolution of institutions for common action*. Cambridge: Cambridge University Press.
Pye-Smith, C., & Feyerabend, G. (1994). *The wealth of communities*. London: Earthscan.
Schelling, T. C. (1978). *Micromotives and macrobehavior*. New York: W.W. Norton & Co.
Schumpeter, J. (1912). *The theory of economic development, translated 1934*. Cambridge, MA: Harvard University Press.
Slobodkin, L. B. (1964). The strategy of evolution. *American Scientist, 52*, 342–357.
Steele, J. (1996). Regime shifts in fisheries management. *Fisheries Research, 25*, 19–23.
Wolinsky, S. M., Korber, B. T. M., Neumann, A. U., Daniels, M., Kunstman, K. J., Whetsell, A. J., et al. (1996). Adaptive evolution of human immunodeficiency virus-type 1 during the natural course of infection. *Science, 272*, 537–542.

Chapter 14
Askö 1995: Commentary by Stephen Carpenter

Stephen Carpenter

14.1 Introduction

Transformation and persistence coexist in all living systems. The concept of resilience was invented to address this paradox (Holling 1973; Folke 2006). Resilience is not about an equilibrium of transformation and persistence. Instead, it explains how transformation and persistence work together. Resilience explains how systems assimilate disturbance, innovation, and change, while simultaneously maintaining characteristic structures and processes (Westley et al. 2006).

In this Askö paper, Levin et al. extend the concept of resilience from its origins in ecology to socioeconomic systems. Originally, resilience was intended to address evolved systems (such as genes and organisms) as well as systems that are self-organized from evolved and abiotic components (such as ecosystems) (Holling 1973; Folke et al. 2004). As societies and economies are self-organized from evolved and abiotic components (Levin 1998), resilience should apply to socioeconomic systems. However, applicability does not guarantee utility. Subsequent research has shown that Levin et al. were correct in anticipating the value of resilience concepts for analysis and management of social–ecological systems (Folke et al. 2004; Gunderson and Folke 2005; Folke 2006; Walker and Salt 2006; Westley et al. 2006).

Levin et al. are particularly interested in using resilience to tackle pressing questions of sustainability. Thus they go well beyond traditional ecological theory, and explore the roles of cooperation, communication, trust and institutional design for resilient and sustainable socioeconomic questions.

S. Carpenter
Center for Limnology, University of Wisconsin, USA
e-mail: srcarpen@facstaff.wisc.edu

T. Söderqvist et al. (eds.), *Bringing Ecologists and Economists Together*,
DOI 10.1007/978-90-481-9476-6_14, © Springer Science+Business Media B.V. 2011

14.2 Scale

Scale is one answer to the paradox of resilience, of transformation and stability working together. Ecologists have found that ecosystem puzzles are often solved by considering the scales at which key processes occur (Levin 1992). Specifically, scale refers to spatial extent of the system, grain of the finest spatial subdivisions of the system, turnover time of major system components, and the highest temporal frequency that is needed to understand the system. A system can be undergoing transformation at certain scales, while maintaining stability at other scales. For example, a patchwork of fires in a forest re-initializes successional changes in the vegetation, while unburnt forest patches contribute seeds from a larger, slower scale, and at the same time microbial dynamics maintain nutrient supplies for growing plants through equilibrial dynamics at a smaller, faster scale. Such patterns across a range of scales often explain resilience of ecosystems, as well as breakdowns of resilience. Management of forests to suppress fire may allow fuel to build up across the forested landscape, leading eventually to fires that are spatially extensive and unusually hot. This removes the seed source from larger scales and burns up the soil and microbes, thereby carrying the entire forest into a completely different regime.

The translation of scale concepts to socioeconomic systems is not simple (Gibson et al. 2000; Westley et al. 2002). Institutions can be scaled by the number of people they affect, and the time horizon of institutional change. Some features of human demography, governance and economics can be mapped to dimensions that resemble ecological scaling. On the other hand, human capacities for abstraction and reflexivity, forward-looking action, and technology development are strikingly different from non-human systems (Westley et al. 2002). These capacities enable human systems to transcend constraints of scale. Thus patterns of change across scales may not provide satisfactory explanations of resilience in socioeconomic systems. For socioeconomic systems, it may be necessary to address cross-system interactions, retaining the many dimensions that differentiate subsystems, instead of compressing the dynamics to one or two scaling dimensions (Westley et al. 2006).

Thus cross-scale, or cross-system, interactions are a critical research frontier for resilience of social–ecological systems. How do multiple subsystems interact to create, or erode, resilience? What kinds of social–ecological configurations lead to sustainable versus unsustainable societies? These are important questions that are just beginning to be addressed.

14.3 Models and Measurements

There is no unique model for resilience. Generally speaking, a model is a concrete representation (such as an equation, a computer program, a physical representation, or a work of art) that can be manipulated to help understand an abstraction (Root-Bernstein 1999). A rich, complex abstraction (like resilience) evokes many models.

In order to use an abstraction for decision or management we must agree on specific, explicit models. We are not restricted to just one model, but we do need to assess the implications of explicit models in transparent ways that are accessible to all. For resilience, modeling is in an early, turbulent (and therefore intellectually exciting) phase (Carpenter et al. 2005). Thus, it is not surprising that the early commentaries on this Askö paper emphasized questions about model choice and measurement.

Since the paper was written, researchers have paid some attention to models and measurement. Any assessment of resilience must answer the question "resilience of what to what?" (Carpenter et al. 2001). The answers to this question imply certain decisions about model details and metrics (Allen et al. 2005; Bennett et al. 2005; Berkes and Seixas 2005; Cumming et al. 2005). The idea of measuring resilience by the size of the stability domain for some specified stable configuration has become more popular among modelers of social–ecological systems. At the time of writing, this seems to be one of the more popular metrics for assessing resilience. The stability-domain approach has led to important insights about resilience in social–ecological systems. It has succeeded in explaining catastrophic change, for example (Scheffer et al. 2001). It has even led to early-warning indicators of collapse (Kleinen et al. 2003; Carpenter and Brock 2006).

As just one example from a rich literature, Scheffer et al. (2003) have used stability domains to investigate the dynamics of policy change in relation to whether societies are tolerant or conformist. When societies are open to new models, tolerate divergent views, and openly share and interpret monitoring data, then adaptive shifts in policy tend to occur smoothly. In contrast, when societies are prejudiced against new models, use sanctions to enforce conformity to established views, and do not openly evaluate monitoring data, policy shifts occur as delayed and massive responses to severe resource crises.

In spite of these successes, the stability domain approach seems to have limitations. Social–ecological systems go through periods of reorganization, in which new ideas become reality. Novelty, in society as in genomes, often results from recombination of existing components, or "memory" of old components that can be reconfigured in new ways. More rarely, innovations are genuinely new. Whether they originate de novo or from reconfiguration of existing elements, the processes that evoke innovation, and then nucleate into new actions, businesses, practices or programs, are essential for the sustainability transition. While innovation is a critical process for addressing sustainability challenges, at the time of writing we have few models and metrics to build upon. How are new stability domains created? This is a frontier of resilience science.

14.4 Theory and Practice

Resilience concepts evolved through the interplay of academic research and management practice (Holling 1986; Gunderson et al. 1995; Folke et al. 2004; Walker and Salt 2006). This evolution is still going on (Walker et al. 2006;

Gunderson and Folke 2005). Centrally-controlled ecosystem management fails because planners view the management as a solution to a problem (Gunderson et al. 1995; Holling and Meffe 1996; Walker and Salt 2006). In contrast, adaptive approaches view management as an opportunity to search for solutions in a situation where uncertainty is high and control points are few. Experienced organizers of adaptive management processes find that resilience concepts are valuable at several levels. For example, resilience concepts help the manager understand the process itself (Gunderson et al. 1995), aid in systems analysis of the problem (Bennett et al. 2005) and open space for collaboration of diverse stakeholders in the management process (Westley et al. 2006).

To apply resilience concepts, they must be communicated outside the boundaries of academia. Communication entails simplification. As simplifications are adopted and harden into rules, application becomes possible but richness and complexity are lost. This hazard seems inescapable. It can be mitigated by ongoing connections of theory and practice, so that both are continually changing in connection with the social–ecological system that we seek to sustain. Thus ongoing coevolution with practice and actual social–ecological systems is also a key research need.

14.5 Conclusions

The Askö paper by Levin et al. represents an important step in the globalization of resilience, by connecting resilience to the broad debate about development, poverty reduction, and sustainability. Folke (2006), in a historical review, explains how resilience concepts have evolved as a consequence of application to social–ecological systems. Concern about global change brings the paradoxical coexistence of transformation and persistence into sharp focus. If society's approaches to nature are not transformed, human well-being will not be sustained (Millennium Ecosystem Assessment 2006). Resilience – transformation and sustainability working together – is the key idea as society searches for changes that will enable us to maintain what we value as the future unfolds.

References

Allen, C. R., Gunderson, L., & Johnson, A. R. (2005). The use of discontinuities and functional groups to assess relative resilience in complex systems. *Ecosystems, 8*, 958–966.

Bennett, E. M., Cumming, G. S., & Peterson, G. D. (2005). A systems model approach to determining resilience surrogates for case studies. *Ecosystems, 8*, 945–957.

Berkes, F., & Seixas, C. S. (2005). Building resilience in lagoon social–ecological systems: A local-level perspective. *Ecosystems, 8*, 967–974.

Carpenter, S. R., & Brock, W. A. (2006). Rising variance: A leading indicator of ecological transition. *Ecological Letters, 9*, 311–318.

Carpenter, S. R., Walker, B. H., Anderies, J. M., & Abel, N. (2001). From metaphor to measurement: Resilience of what to what? *Ecosystems, 4*, 765–781.

Carpenter, S. R., Westley, F., & Turner, M. G. (2005). Surrogates for resilience of social–ecological systems. *Ecosystems, 8*, 941–944.
Cumming, G. S., Barnes, G., Perz, S., Schmink, M., Seiving, K. E., Southworth, J., et al. (2005). An exploratory framework for the empirical measurement of resilience. *Ecosystems, 8*, 975–987.
Folke, C. (2006). Resilience: The emergence of a perspective for social-ecological systems analyses. *Global Environmental Change, 16*, 253–267.
Folke, C., Carpenter, S. R., Walker, B. H., Scheffer, M., Elmqvist, T., Gunderson, L. H., et al. (2004). Regime shifts, resilience and biodiversity in ecosystem management. *Annual Review in Ecology, Evolution and Systematics, 35*, 557–581.
Gibson, C. C., Ostrom, E., & Ahn, T. K. (2000). The concept of scale and the human dimensions of global change: A survey. *Ecological Economics, 32*, 217–239.
Gunderson, L., & Folke, C. (2005). Resilience – Now more than ever. *Ecology and Society, 10*(2), 22. [online] URL: http://www.ecologyandsociety.org/vol10/iss2/art22/.
Gunderson, L. H., Holling, C. S., & Light, S. S. (Eds.). (1995). *Barriers and bridges to the renewal of ecosystems and institutions*. New York: Columbia University Press.
Holling, C. S. (1973). Resilience and stability of ecological systems. *Annual Review of Ecology and Systematics, 4*, 1–23.
Holling, C. S. (1986). The resilience of terrestrial ecosystems: Local surprise and global change. In Clark W. C. & Munn R. E. (Eds.), *Sustainable development of the biosphere* (pp. 292–317). London: Cambridge University Press.
Holling, C. S., & Meffe, G. K. (1996). Command and control and the pathology of natural resource management. *Conservation Biology, 10*, 328–337.
Kleinen, T., Held, H., & Petschel-Held, G. (2003). The potential role of spectral properties in detecting thresholds in the earth system: Application to the thermohaline circulation. *Ocean Dynamics, 53*, 53–63.
Levin, S. A. (1992). The problem of pattern and scale in ecology. *Ecology, 73*, 1943–1967.
Levin, S. A. (1998). Ecosystems and the biosphere as complex adaptive systems. *Ecosystems, 1*, 431–436.
Millennium Ecosystem Assessment. (2006). *Ecosystems and human well-being: General synthesis*. Washington, DC: Island Press.
Root-Bernstein, R., & Root-Bernstein, M. (1999). *Sparks of genius*. Boston: Houghton-Mifflin USA.
Scheffer, M., Carpenter, S. R., Foley, J. A., Folke, C., & Walker, B. H. (2001). Catastrophic shifts in ecosystems. *Nature, 413*, 591–596.
Scheffer, M., Westley, F., & Brock, W. (2003). Slow response of societies to new problems: Causes and costs. *Ecosystems, 6*, 493–502.
Walker, B., Gunderson, L., Kinzig, A., Folke, C., Carpenter, S., & Schultz, L. (2006). A handful of heuristics and some propositions for understanding resilience in social-ecological systems. *Ecology and Society, 11*(1), 13. [online] URL: http://www.ecologyandsociety.org/vol11/iss1/art13/.
Walker, B., & Salt, D. (2006). *Resilience thinking: Creating space in a shrinking world.* Washington, DC: Island Press.
Westley, F., Carpenter, S. R., Brock, W. A., Holling, C. S., & Gunderson, L. H. (2002). Why systems of people and nature are not just social and ecological systems. In L. H. Gunderson & C. S. Holling (Eds.), *Panarchy: Understanding transformations in human and natural systems* (pp. 103–120). Washington, DC: Island Press.
Westley, F., Zimmerman, B., & Patton, M. (2006). *Getting to maybe*. Toronto, Canada: Random House.

Chapter 15
Askö 1995: Commentary by David A. Starrett

David A. Starrett

The concepts of resilience and sustainability have been prominent in the popular press in recent years, notably in connection with the Millennium Ecosystem Assessment released in 2005. They are central to the debate concerning how to frame sound environmental policy but are difficult to define precisely, much less measure accurately. Policies lead to sustainability if they enable some measure of well-being such as the "standard of living" to be perpetuated into the indefinite future. The system is resilient if it is somehow able to right itself in the face of random or manmade disturbances in a sustainable way. But the questions of what exactly needs to be sustained and how to measure resilience and determine whether a system is resilient has caused much confusion in the past.

This paper by Levin et al. was one of the first to give a scientific discussion of the resilience concept and helped to set the framework for the ambitious Millennium Ecosystem Assessment that followed. It was written by a group of economists and ecologists, a selection of which have been meeting under the auspices of the Beijer Institute for some years. These groupings might be seen as a microcosm of the much more comprehensive Millennium Ecosystem Assessment, which involved the work of many scientists from a wide range of disciplines. However, because the Beijer meetings are at a much smaller scale they allow much more personal interaction and fully exploit the synergies across disciplines.

Economists come to the table with a lot of experience analyzing growth models involving a small number of variables and have established criteria for sustainability in these. They also have studied in depth the concept of stability, namely the tendency of the system to return to an "equilibrium" state after some outside perturbation. In many of these models, simple stability measures could serve to measure resilience. However, they were aware that if there were sufficient nonlinearities in the system, systems that were stable (and thus resilient) in the face of small perturbations might be quite unstable in the face of large perturbations.

D.A. Starrett
Department of Economics, Stanford University, USA
e-mail: davstarret@aol.com

T. Söderqvist et al. (eds.), *Bringing Ecologists and Economists Together*,
DOI 10.1007/978-90-481-9476-6_15, © Springer Science+Business Media B.V. 2011

This phenomenon emerged in models where there was a destabilizing feedback mechanism. The best known of these was the "poverty trap" wherein once a country becomes poor enough, subsistence requires consuming almost all resources, reduces investment and thereby productive capacity leading to a downward spiral from which there is no obvious escape.

Ecologists are used to studying the dynamics of models with many interacting variables and significantly nonlinear dynamics. Such models are sufficiently complicated that one can rarely solve them completely to determine the long run growth path. However they demonstrate many instances of multiple basins of attractions and sudden shifts among these. They have seen many instances of destabilizing positive feedback, perhaps the best know of which is the fisheries collapse model wherein once fishing effort reduces stocks to a critical level, reproduction in impaired and even if fishing is reduced the stock spirals downward.

Thus by bringing the economics and ecology traditions together, it was possible to refine the concept of resilience. It emerged that resilience is a very complicated concept, difficult to measure and hard to guarantee in systems where some aspects of the dynamics are unknown or uncertain. The reason is that without knowing the precise nature of system dynamics one cannot be sure whether or not small perturbations of the system might lead to large (and possibly disastrous) shifts in the state of that system.

In the aftermath of the resilience meeting, similar groups of economists and ecologists have refined their thinking and developed and studied in detail a simple one-state-variable "Paradigm" model that illustrates many of the complex behaviors that more complicated systems exhibit. It is known as the "shallow lake" model. (This model and its properties and history are discussed in a special issue of *Environment and Resource Economics*, volume 26, 2003). The situation is one in which phosphorus is introduced into a relatively shallow lake through fertilizer runoff from agriculture. The destabilizing feedback derives from the fact that whereas the lake bed can store phosphorus for some time, if the loading is too high it loses that capacity, releases phosphorus back to the water, leading to increased turbidity through algae blooms. This model can exhibit sudden (catastrophic) shifts between clean and dirty states, and depending on parameters can become trapped in the dirty state forever even if phosphorus loading is reduced to zero.

The main policy implications that emerge from these studies is that in our linked socioeconomic and ecological systems, management needs to be especially vigilant in monitoring system health and cautious in allowing large perturbations with unknown consequences. Further, institutions must be flexible in order to deal quickly with unexpected changes. These warnings were voiced clearly in the resilience article and have been echoed since in other forums such as the aforementioned Millennium Ecosystem Assessment. We can only hope that the world's governments take heed in a timely manner.

Chapter 16
Askö 1997: Food Production, Population Growth, and the Environment*

Gretchen Daily, Partha Dasgupta, Bert Bolin, Pierre Crosson, Jacques du Guerny, Paul Ehrlich, Carl Folke, AnnMari Jansson, Bengt-Owe Jansson, Nils Kautsky, Ann Kinzig, Simon A. Levin, Karl-Göran Mäler, Per Pinstrup-Andersen, Domenico Siniscalco, and Brian Walker

* This article was published in *Science*, Vol. 281. Daily et al., Food production, population growth. and the environment, 1291–1292, Copyright AAAS (1998).

G. Daily and P. Ehrlich
Department of Biological Sciences, Stanford University, Stanford, CA 994305, USA

P. Dasgupta
Faculty of Economics and Politics, University of Cambridge, Austin Robinson Building, Sidgwick Avenue, Cambridge CB3 9DD, UK; Beijer Institute of Ecological Economics, Stockholm, Sweden; Resources for the Future, Washington DC, USA

B. Bolin
Department of Meteorology, University of Stockholm, 10691 Stockholm, Sweden

P. Crosson
Resources for the Future, 1616 P Street, Washington, DC 20036, USA

J. du Guerny
Food and Agriculture Organization, Rome, Italy

C. Folke, A. Jansson, and N. Kautsky
Department of Systems Ecology, University of Stockholm; The Beijer Institute of Ecological Economics, Stockholm, Sweden

B.-O. Jansson
Department of Systems Ecology, University of Stockholm, Sweden

A. Kinzig
Princeton Environmental Institute, Princeton University, Princeton, NJ 08544, USA

S.A. Levin
Department of Ecology and Evolutionary Biology, Princeton University, Princeton, NJ 08544, USA

K.-G. Mäler
The Beijer Institute of Ecological Economics, Stockholm, Sweden; Stockholm School of Economics, Stockholm, Sweden

P. Pinstrup-Andersen
International Food Policy Research Institute, Washington, DC 20006-1002, USA

D. Siniscalco
Fondazione Eni Enrico Mattei, Milan 20123, Italy

B. Walker
CSIRO Lyneham, ACT, Australia

T. Söderqvist et al. (eds.), *Bringing Ecologists and Economists Together*,
DOI 10.1007/978-90-481-9476-6_16, © Springer Science+Business Media B.V. 2011

There are two broad criteria by which one can judge humanity's success in feeding itself: (i) the proportion of people whose access to basic nutritional requirements is secure, and (ii) the extent to which global food production is sustainable. Even though the two are related, they have usually been discussed separately in popular writings. This has had unfortunate consequences. Writings on the sustainability of the food supply have often encouraged readers to adopt an all-or-nothing position (predicting a rosy or a catastrophic future), which has drawn attention from the economic misery endemic in large parts of the world. On the other hand, writings on the adequacy of the world's food supply frequently conclude with the truism that the nearly one billion people in poor countries who go to bed hungry each night do so because they are extremely poor. In short, the second "sustainability" approach has focused on aggregate food production and its future, whereas the first has isolated food-distribution failure as a cause of world hunger. Here we argue that these two questions should not be studied separately, that their link is revealed in the interactions between ecological and economic systems operating primarily at the geographically localized level, and that policy interventions must target this link.

16.1 Global Sketches

World population has increased at a historically high average annual rate of 1.8% since 1950. However, cereal production (accounting for more than 50% of the energy intake of the world's poor today) has more than kept pace: It has increased from 275 kg per person in the early 1950s to 370 kg per person in the early 1980s. This change has been accompanied by improvements in a number of indicators of human well-being, such as gross output per head, infant mortality rate, life expectancy at birth, and literacy. Much of the complacency economists have displayed in recent years about food availability can be traced to these recorded improvements.[1]

The problem with this complacency is that conventional indicators of the standard of living pertain to commodity production, not to the natural resource base on which all production depends. These indicators cannot say whether, for example, increases in gross national product (GNP) per head are being realized by a depletion of natural capital; in particular, whether increases in agricultural production are being achieved by a "mining" of soil, lowering of water tables, and impairment of other ecosystem services. Such impairment can easily go unrecorded, because the use of ecosystem services all too often involves transactions that are not mediated by an effective price system. So, for example, if individual farmers, when drawing water from an aquifer, ignore the fact that their extraction will increase others' future extraction costs because of a lowering of the water table, the social cost of agricultural production would exceed the farmers' private costs. Even though each farmer would typically impose only a small additional cost on others, the sum of the costs imposed by each on all others could well be substantial. This means that the real costs of agricultural production could exceed the market prices of agricultural produce. Indeed, their

[1] Two recent examples: Simon (1996, 1997).

market prices could decline over time while the real cost of production rose (see the figure, where a hypothetical case is shown). By concentrating on current welfare measures such as GNP, market prices of agricultural produce, and life expectancy at birth, economists, journalists, and political leaders have for the most part wrongly bypassed the links that exist between population growth, increased material output, and the state of the natural resource base.

Ecologists' findings suggest that a near-50% increase in world population, allied with a doubling of gross world product per head, would by 2030 create substantial additional stresses in both local and global ecosystems (Vitousek et al. 1986, 1997a). Global demand for food could easily double over the period 1990–2030, with two-and-a-half- to threefold increases in the poorest countries. Of particular concern are Asia and Africa where, over the next 50 years, plant-derived food energy requirements are expected to increase by a factor of 2.3 and 5, respectively, with a more-than-sevenfold increase expected in some countries (Pinstrup-Andersen 1994; Crosson and Anderson 1995; Food and Agriculture Organization FAO 1996).

Such increases in food requirements mean that we must manage constraints on the supplies of production inputs and on the environmental consequences of the use of these inputs. Increases in food production will in great measure have to continue to come from increased yields from land already in production and from improved efficiency in the use of water already co-opted by agriculture.[2]

But there are obstacles here. First, working against the trend of increased production from transgenic crops, many of the genetic resources required for the development

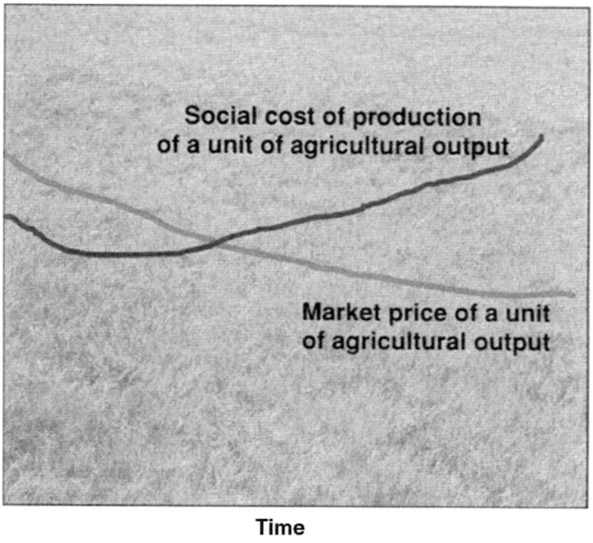

Hidden costs. A hypothetical example showing that apparent decreases in the market price of a commodity can be accompanied by an increase in the actual production cost if the social cost of the commodity is rising

[2]The prospects of seafood production appear to be severely limited (Safina 1995; Folke et al. 1998).

of such improved crops are being destroyed (Waterlow et al. in press; FAO 1995; Daily 1997; Postel et al. 1996). Second, a number of the ecosystem services underpinning productivity are under increasing threat (Holdren and Ehrlich 1974; Daily 1995). Third, the environmental and human health consequences of fertilizer and pesticide application in intensive production systems are of growing concern (Vitousek et al. 1997b; Matson et al. 1998). And fourth, extreme climatic events accompanying global climate change may be expected to present an additional threat to food security; for example, to pest control and crop yields (Bruce et al. 1996).

16.2 Food Viewed Through a Local Lens

These obstacles are not present uniformly across the globe. Moreover, local problems of production and distribution can be difficult to counter even when global supplies are adequate, because the purchasing power of the poor is weak. To ask merely whether global food supplies can be sustainably increased to meet future requirements misses much of the question. Food scarcity manifests itself locally, so efforts to alleviate it must be tailored to local circumstances. To do otherwise is akin to doctoring a sick person on the basis of global health statistics.

Correct diagnosis of the problems at the population–food–environment nexus is usually a local matter, even though appropriate treatment may require regional and global support. For example, soil erosion may not currently be a serious threat to global agricultural capacity, but locally in various parts of the world it presents major problems to the people affected (see, for example, Pimentel et al. 1995; Crosson 1995, and references therein). Similarly, decisions concerning fertility, education, child care, food, work, health care, and the use of the local natural resource base are in large measure reached and implemented within households, which face constraints that are shaped in part by national and international policy. The influence of household decisions is felt through local interactions (for example, intra-village and village–town trades) and thence upward globally. Recent work has identified a variety of circumstances in which there exist positive, synergistic feedback mechanisms driving poverty, hunger, fertility, environmental degradation, and civic disconnection at the local level, even while national (and not merely global) income is rising (Dasgupta and Mäler 1991; Dasgupta 1993, 1995; Cleaver and Schreiber 1994). This suggests that, if we are to obtain a reliable projection of global food prospects, we need to adopt a local, contemporary lens.

16.3 Moving into a More Nonlinear Domain?

Such interactions as we have alluded to above may well be a signal that local food systems in many places are moving into a more nonlinear domain, in which local environmental pressures exerted by growing populations rise more than in proportion to the growth, other things being equal. Moreover, in agriculture, for example, the consequences of a given shock today to the production system would likely be proportionately greater than in earlier decades, for at least three reasons. First, the food system is losing genetic heterogeneity. Thus, the vulnerability of crops to atypical weather events or pest outbreaks will probably increase. Second, the world is becoming more tightly coupled through globalization, so that, in the absence of adequate international insurance arrangements, global fluctuations in food production could well have stronger effects on local food availabilities. Third, increased intensity of crop and livestock production and centralization of food processing leaves large regions vulnerable to disease outbreak and to reduced efficacy of antibiotics (Witte 1998).

16.4 Policies

Threats to environmental security very often come allied to institutional failure. Thus, when thinking about environmental security, particular attention needs to be given to the institutions in which individuals, households, firms, and communities go about their business. Evidence suggests that open societies harboring secure property rights (be they private or communal) and avoiding flagrantly distortionary fiscal policies are not only desirable in themselves, they would also appear to be good for the sustainable management of the natural resource base.

One aspect of such institutional reform should be increased public investment in agriculture and in people in poor countries. Such investment needs to be directed at new technologies and institutions at the local level involving, among other things, health care, reproductive health, and education (the latter, when benefiting women, has been found to be an important determinant of fertility behavior). Evidence also directs attention to the importance of dissemination of technical knowledge to local populations and to the need for local populations to be in a position to adapt new knowledge to their particular circumstances. Greater public investment needs to be made in the global biophysical, hydrological, and institutional assets that are crucial to increasing yield – water-use efficiency; gene banks; local land races; and protection of wild crop relatives and biodiversity generally, both for genetic material and for ecosystem services.

It would be desirable also to establish local "foresight institutions," whose purpose would be to monitor key local trends in aspects of food production to inform local and global policy. Fine-tuning a system as complex as the food system and capitalizing on the potential for institutional changes to boost food production in a sustainable fashion will require much more detailed local information than is typically available today. For instance, trends in the effects of soil degradation on land productivity are extremely important but not well understood, and so are vigorously disputed (Pimentel et al. 1995; Crosson 1995). Similarly, the geographic occurrence and security of dwindling populations of wild crop relatives is poorly known.

Translation of data into a workable set of social indicators is also important. Such indicators enable policy debate to be conducted in an illuminating manner.

In this context, "green accounting" and genuine progress indicators are needed. Developing the local capacity to collect and distribute information is integral to any sustainable development program.

References

Bruce, J. P., Lee, H., & Haites, E. F., (Eds.) (1996). *Climate change 1995: Economic and social dimensions of climate change. Contribution of Working Group III to the Second Assessment Report of the Intergovernmental Panel on Climate Change*. Cambridge: Cambridge University Press.

Cleaver, K. M., & Schreiber, G. A. (1994). *Reversing the spiral: The population, agriculture, and environment nexus in Sub-Saharan Africa*. Washington, DC: World Bank.

Crosson, P. (1995). Soil erosion estimates and costs. *Science*, 269, 461–464.

Crosson, P., & Anderson, J. R. (1995). Demand and supply: Trends in global agriculture. *Food Policy, 19*, 105–119.

Daily, G. C. (1995). Restoring value to the world's degraded lands. *Science, 269*, 350–354.

Daily, G. C. (Ed.) (1997). *Nature's services: Society's dependence on natural ecosystems.* Washington, DC: Island Press.

Dasgupta, P. (1993). *An inquiry into well-being and destitution.* Oxford, UK: Clarendon Press.

Dasgupta, P., & Mäler, K.-G. (1991). *The environment and emerging development issues.* Proceedings of the Annual Bank Conference on Development Economics, 1990. Supplement to the World Bank Economic Review, 1991. Washington DC: The World Bank.

Dasgupta, P. (1995). The population problem: Theory and evidence. *Journal of Economic Literature, 33*, 1879–1902.

FAO. (1995). *World food survey*. Rome: FAO.

Folke, C., Kautsky, N., Berg, H., Jansson, Å., & Troell, M. (1998). The ecological footprint concept for sustainable seafood production: A review. *Ecological Applications, 8*, 63–71.

Food and Agriculture Organization (FAO). (1996). *World food summit technical document 4.* Rome: FAO.

Holdren, J. P., & Ehrlich, P. R. (1974). Human population and the global environment. *American Scientist 62*, 282–292.

Matson, P. A., Naylor, R., & Oritz-Monasterio, I. (1998). Integration of environmental, agronomic, and economic aspects of fertilizer management. *Science, 280*, 112–115.

Pimentel, D., et al. (1995). Environmental and economic costs of soil erosion and conservation benefits. *Science, 267*, 1117–1123.

Pinstrup-Andersen, P. (1994). *World food trends and future food security*. Food Policy Report. Washington, DC: International Food Policy Research Institute.

Postel, S. L., Daily, G. C., & Ehrlich, P. R. (1996). Human appropriation of renewable fresh water. *Science, 271*, 785–788.

Safina, C. (1995). The World imperiled fish. *Scientific American, 273*, 46.

Simon, J., (1996). *The ultimate resource* 2. Princeton University Press, Princeton NJ.

Simon, J. (1997). *Economist, 345*, 21.

Vitousek, P. M., Ehrlich, A., Ehrlich, P., & Matson, P. (1986). Human appropriation of the products of photosynthesis. *Bioscience, 36*, 368–373.

Vitousek, P. M., Ehrlich, A., Ehrlich, P., & Matson, P. (1997a). *Symposium on the scale of human activity*. Science, 20 July 1997.

Vitousek, P. M., et al. (1997b). Human alteration of the global nitrogen cycle: Sources and consequences. *Ecological Applications, 7*, 737–750.

Waterlow, J. C., Armstrong, D. G., Fowden, L., & Riley, R. (Eds.). (in press). *Feeding a World population of more than eight billion people: A challenge to science.* New York: Oxford University Press.

Witte, W. (1998). Medical consequences of antibiotic use in agriculture. *Science* 279, 996–997.

Chapter 17
Askö 1997: Commentary by Robert J. Scholes

Robert J. Scholes

17.1 We Live in a Remarkable Period

For the first time since the Neolithic agricultural revolution about 7,000 years ago, a turning point in the apparently inexorable growth of the human population is in sight (Lutz et al. 2001). The mean number of children per female has dropped below replacement levels in most developed countries, and will do so in most developing countries in the next 50 years. This is not for 'Malthusian' density-dependent reasons such as resource limitation, rampant disease, starvation and conflict, but rather because of rising average human wellbeing and increasing urbanization. But before we congratulate ourselves on having apparently defused the population bomb, we need to consider the challenges and implications of feeding the additional approximately three billion people that demographic momentum commits us to by 2050 (UN 2005). This issue is the point of departure of Daily et al.

Daily et al. argue that there are likely to be nasty surprises built into the coupled human–agriculture system, largely the result of our failure to fully account for the costs of agriculture. In particular, decreases in 'natural capital' are not reflected in the price of agricultural products, and the consequences of declining natural capital are not factored into optimistic projections of future agricultural production. As a result, we could either fail to meet the mid-twenty-first century food security challenge, or if we do, it may be unsustainable because in the process we have undermined the ecological basis for human wellbeing: The continued adequate supply of a variety of essential ecosystem services, including food itself, but also usable water, a habitable climate and the genetic variation necessary for future adaptation.

R.J. Scholes
Council for Scientific and Industrial Research, PO Box 395, Pretoria 0001, South Africa
e-mail: bscholes@csir.co.za

T. Söderqvist et al. (eds.), *Bringing Ecologists and Economists Together*,
DOI 10.1007/978-90-481-9476-6_17, © Springer Science+Business Media B.V. 2011

17.2 Accelerating Impacts with Delayed Consequences

While most of the potentially-negative effects of agriculture on natural resources (for instance, soil erosion, nutrient depletion and overabundance, habitat loss and the emergence of pests) are as old as farming itself, the package of technologies that comprise 'modern agriculture' is barely 50 years old, and we do not have a clear understanding of their long-term sustainability. Both the breadth and depth of the agricultural footprint increased dramatically in the twentieth century, during which mechanization, synthetic fertilizers, novel pesticides, classical plant breeding and pumped irrigation were implemented at large scale. These technologies enabled the post-1960 'green revolution', which allowed agricultural production to increase at a greater rate than population, giving hope of a food-secure world in the future. But the impact of food production and harvesting on the biosphere also accelerated, and since its consequences has a built-in time lag of several decades, so it is far too soon to declare victory. For instance, the post-1960 acceleration in greenhouse gas emissions is noticeable, as is the rise in global temperatures and sea levels. Despite most of the increased food production coming from greater yields per unit farmed land, the cultivated area has increased substantially since 1960, and biodiversity continues to fall (MA 2005).

Most of these factors have negative feedbacks on the level or reliability of ecosystem services in the medium term, including on food production itself. The accelerating trends in the direct drivers give credence to concern expressed by Daily et al. that we may be entering a period of nonlinearity, with potential for abrupt and irreversible change. Will such changes first manifest themselves in the 'centre', i.e. in high-potential places where agriculture has been most intense, for the longest period of time, or will they emerge on the 'periphery', the increasingly-marginal areas where livelihoods are precarious and the options for ecosystem service substitution are few? While there are some worrying syndromes in the 'centre' (such as the spreading 'dead zones' in coastal waters), I suspect that the least-manageable problems will be at the margins. Self-propagating failures will emerge as localized collapses that progressively unravel through spatial knock-on effects, such as the leakage of sediment and nutrients, spread of diseases and migration of people in search of viable livelihoods. In all likelihood, these signals will originate in the poor, populous and arid parts of the world, physically and conceptually distant from the temperate, wealthy regions where global decisions are made.

17.3 What Can Be Done?

Several years after publication the Daily et al. article, its basic premises remain valid, but progress on the six specific policy suggestions remains patchy. It is questionable whether global society is becoming more 'open', either in terms of fair trade or in terms of secure property rights. The World Summit on Sustainable Development in

Johannesburg in 2002 focused attention on the collective need to address poverty, but progress towards the Millennium Development Goals is mixed – with Africa, in particular, falling behind, and very few developed countries meeting their Official Development Aid commitments. The dissemination of global technical knowledge to local level, and the financial support of public goods management remain major challenges in practice, despite many well-intentioned words spoken about them. The funding of agricultural public goods institutions such as the Consultative Group on International Agricultural Research (CGIAR) continues to stagnate. Progress has been made on Assessments: The Millennium Ecosystem Assessment (MA 2005), and the International Assessment of Agricultural Science and Technology for Development are directly relevant to this topic. The MA explicitly linked food production to other aspects of the ecosystem services that collectively comprise 'natural capital', as advocated in the Daily et al. paper, and pioneered the notion of integrated assessments at multiple scales, from the global to local. The holy grail of an agreed, sensitive and workable set of social indicators remains work in progress.

Over and above to the policy actions suggested by Daily et al., there needs to be an informed and deliberate re-balancing of the often conflicting imperatives of short-term yield maximization and long-term resilience. For instance, in the presence of cheap fertilizer and agricultural production subsidies, many farmers have maximized yield her hectare by fertilizing to the point where there is no further yield response. However, at this 'local optimum' point, there is substantial nutrient leakage to water and air, with costly off-site impacts. Fertilization at levels that have much reduced leakage still allows 80% of the maximum yield to be achieved. At the other end of the spectrum, most farming systems in hunger-beset sub-Saharan Africa are chronically under-fertilized, and a strong argument can be made that moderate agricultural intensification will be better overall for biodiversity and ecosystem service maintenance as well as food production and human nutrition than massive extensification. Similar intelligent trade-off decisions could be made around most agricultural technologies if better information were available about their impacts on natural capital at all scales. That is the essence of the plea in the Daily et al. paper. We have made some progress towards that goal, but not nearly enough.

References

Lutz, W., Sanderson, W., & Scherbov, S. (2001). The end of world population growth. *Nature, 412*, 543–545.

MA. (2005). *Ecosystems and human wellbeing: Our human planet. Summary for decision makers.* Washington, DC: Millennium Ecosystem Assessment and Island Press.

UN. (2005). *World population prospects: The 2004 revision.* New York: United Nations Population Division.

Chapter 18
Askö 1997: Comment by Anne-Sophie Crépin

Anne-Sophie Crépin

18.1 More Reasons for Concern?

Has humanity become more successful in feeding itself than it was when Daily et al. published their article? Since then world population has increased with nearly one billion people and even if its growth seems to be ending sometimes during this century, we will still most probably have to feed two more billion people by 2050 (Lutz et al. 2008). Meanwhile recent global assessments and research witness that the planet is put under increasing pressure. (Steffen et al. 2004; Millennium Ecosystem Assessment 2005; IPCC 2007; Rockström et al. 2009.) Humans are becoming such a dominating species that we have entered a new geologic era, the anthropocene (Crutzen and Stoermer 2000).

In this context it is not surprising if resource use trade-offs are magnified and the food issue is no exception. Recent developments on food markets illustrate these trade-offs. The supply side could not match the recent rapid increases in demand for food products, which lead to 2009 being a devastating year for the world's hungry, marking a significant worsening of the already bad trend in global food security. The demand for food products is steadily increasing due to population growth but also, which was maybe less expected, to an increased demand for agricultural crops for bio-fuel production. (SOFA 2008; SOFI 2009.)

On the supply side, the situation is worrying as well. The prospects for increases in land-based food production still depend on improving the use of existing agricultural land. The four obstacles to that, which Daily et al. identified still remain: (1) biodiversity depletion, (2) increasing threat to ecosystem services, (3) negative environmental and human health effects of pesticides and fertilizers and (4) extreme climatic events due to climate change. In addition, the climatic threat has become more urgent (Smith et al. 2009) and may lead to a decrease in the amount of cultivated land due to sea level rise, or desertification. (Millennium Ecosystem

A.-S. Crépin
Beijer Institute of Ecological Economics, the Royal Swedish Academy of Sciences, Box 50005, SE-10405 Stockholm, Sweden
e-mail: annesophie.crepin@beijer.kva.se

T. Söderqvist et al. (eds.), *Bringing Ecologists and Economists Together*,
DOI 10.1007/978-90-481-9476-6_18, © Springer Science+Business Media B.V. 2011

Assessment 2005; IPCC 2007.) Furthermore their concern about ecosystems reaching more nonlinear domains has been confirmed for many systems. (See e.g. Resilience Alliance and Santa Fe Institute 2004.)

Daily et al. did not really consider the potential for food production from the seas, which has steadily increased during the second half of the twentieth century. The overall trend for fish stocks is quite depressing though (see Myers and Worm 2003; Worm et al. 2006, and comments by Murawski et al. 2007), and the prospects for increased yields from fisheries seem small unless drastic measures are taken to better manage the existing threatened stocks. Meanwhile aquaculture has been developing steadily. It is the fastest growing animal food-producing sector, now producing half of the fish consumed by world population. However, large parts of the aquaculture industry compete with poor people's demand of small pelagic fish. (Tacon and Metian 2009.) In addition, public concerns about aquaculture practices and fish quality have recently contributed to slower growth rates. Integrated multitrophic aquaculture, which promotes economic and environmental sustainability and organic aquaculture are on the rise to respond to these concerns. (SOFIA 2008.)

The overall picture is not so optimistic: even if there are potential for increase in food production in the short run (Dyson 1999), the prospects for a long, lasting and sustainable increase in food production are quite low due to negative environmental and health effects.

In addition local problems of food production and distribution are still difficult to get rid of even with adequate global supplies. Although widespread in certain regions of the world, food scarcity remains a local phenomenon largely affected by soil quality, local weather events, and potential for using local natural resource but also social aspects like fertility, education, child care, work, and health care availability. Recent evidence also show the massive impact of the HIV/AIDS epidemic in Africa on sustainable food production potential (Gillespie and Kadiyala 2005; Ternström 2005) adding to already existing poverty trap issues.

Paradoxically, the world is simultaneously experiencing a global obesity epidemic (more than one billion adults are overweight) due to increased consumption of energy-dense foods high in saturated fats and sugars, and reduced physical activity (James et al. 2001; WHO 2009).

Problems associated with the food production processes have recently been highlighted as well. Besides the usual negative health and environmental effects due to pesticides and chemical fertilizers use on agricultural land, the climate impacts of the food production systems have come more into focus (see e.g. Weber and Matthews 2008) as well as the increasing use of additives – to get food look nice and conserve better – and their negative health effects. (See e.g. McCann et al. 2007; Food Standards Australia New Zealand 2007; Nilsson 2008.)

18.2 So What Can We Do?

The main points in Daily et al. are more valid than ever. Local interactions between ecological and economic systems create strong links between the issues of food security and sustainable food production at a global level. The only way to get a complete picture of humanity's potential to feed itself is to study security, safety and sustainability issues together. How can we provide food security, safety and a sustainable food production system in a world with a limited amount of arable land, depleted fisheries, rising sea levels and a warming climate?

A first step is to get a realistic picture of the situation. Unfortunately, conventional indicators of living standard that bypass the links between population growth, increased material output and the state of the natural resource base still dominate (GDP, market prices of agricultural products, life expectancy). Meanwhile serious work has been and continues to be done to produce new indicators and test them in practice (see e.g. Markandya and Tamborra 2005; Mäler 2007; Nourry 2008).

Recent development show that Daily et al.'s policy recommendations remain valid and the 2009 prize in economic sciences in memory of Alfred Nobel to Elinor Ostrom recognizes the role of institutions harboring secure property rights (private or communal) for good and sustainable management of the natural resource base. Daily et al. also pointed to the necessity to avoid flagrant distortionary taxes, increase investment in agriculture and in people in poor countries. In this context much remains to be done. In particular we should learn from the mistakes related to the Common Agricultural Policy in the EU, which was certainly successful in providing food security and support people employed in the agricultural sector in post World War II Europe but is failing on the sustainability issue.

It seems obvious that a successful policy will need to treat the problem simultaneously on the demand and on the supply side to mitigate population growth in the long run and improve the sustainability of the food production system. In addition to a need for better institutions and investment on a global scale, support should be given for institutions (health care, education) and new technology at the local level and designed adaptively to help (see e.g. Enfors and Gordon 2008).

References

Crutzen, P. J., & Stoermer, E. F. (2000). The 'Anthropocene'. *Global Change Newsletter, 41*, 17–18.

Dyson, T. (1999). World food trends and prospects to 2025. *Proceedings of the National Academy of Sciences (USA), 96*, 5929–5936. Colloquium Paper.

Enfors, E. I., & Gordon, L. J. (2008). Dealing with drought: The challenge of using water system technologies to break dryland poverty traps. *Global Environmental Change, 18*, 607–616.

Food Standards Australia New Zealand. (2007). Choosing the right stuff – The official shoppers' guide to food additives and labels, kilojoules and fat content. http://www.foodstandards.gov.au/newsroom/publications/choosingtherightstuff/. Retrieved 23 October 2009.

Gillespie, S., & Kadiyala, S. (2005). *HIV/AIDS and food and nutrition security: From evidence to action. Food Policy Review 7*. Washington, DC: International Food Policy Research Institute.

IPCC. (2007). Climate change: Synthesis report, In R. K. Pachauri & Reisinger A. (Eds.) *Contribution of Working Groups I, II and III to the Fourth Assessment. Report of the Intergovernmental Panel on Climate Change, Core Writing Team*. (104 pp.) Geneva, Switzerland: IPCC.

James, P. T., Leach, R., Kalamara, E., & Shayeghi, M. (2001). The worldwide obesity epidemic. *Obesity Research, 9* (Suppl. 4).

Lutz, W., Sanderson, W., & Scherbov, S. (2008). The coming acceleration of global population ageing. *Nature, 451*, 716–718.

Mäler, K.-G. (2007). Wealth and sustainable development: The role of David Pearce. *Environmental and Resource Economics, 37*, 63–75.

Markandya, A., & Tamborra, M., Eds. (2005). *Green accounting in Europe – A comparative study, Vol. 2*. London: Edward Elgar.

McCann, D. A., Barrett, A., Cooper, D., Crumpler, L., Dalen, K., Grimshaw, E., et al. (2007). Food additives and hyperactive behaviour in 3-year-old and 8/9-year-old children in the community: A randomised, double-blinded, placebo-controlled trial. *Lancet, 370*(9598), 1560–1567.

Millennium Ecosystem Assessment. (2005). *Ecosystems and human wellbeing: Our human planet. Summary for decision makers*. Washington, DC: Millennium Ecosystem Assessment and Island Press.

Murawski, S., Methot, R., & Tromble, G. (2007). Biodiversity loss in the ocean: How bad is it? And other responses to Worm et al. (2006). *Science, 316*, 1281–1284.

Nilsson, M.-E. (2008). *Den hemlige kocken: Det okända fusket med maten på din tallrik*. Sweden: Ordfront Förlag AB.

Nourry, M. (2008). Measuring sustainable development: Some empirical evidence for France from eight alternative indicators. *Ecological Economics, 67*, 441–456.

Resilience Alliance and Santa Fe Institute. (2004). Thresholds and alternate states in ecological and social–ecological systems. Resilience Alliance. (online) URL: http://www.resalliance.org/index.php?id=183.

Rockström, J., Steffen, W., et al. (2009). A safe operating space for humanity. *Nature, 461*, 472–475.

Smith, J. B., Schneider, S. H., Oppenheimer, M., Yohe, G. W., Hare, W., Mastrandrea, M. D., et al. (2009). Assessing dangerous climate change through an update of the Intergovernmental Panel on Climate Change (IPCC) 'Reasons for Concern'. *Proceedings of the National Academy of Sciences, 106*, 4133–4137.

SOFA. (2008). *The state of food and agriculture – Biofuels: prospects, risks and opportunities*. Rome: Food and Agriculture Organization of the United Nations (FAO).

SOFI. (2009). *The state of food insecurity in the world 2009 – Economic crises impacts and lessons learned*. Rome: Food and Agriculture Organization of the United Nations (FAO).

SOFIA. (2008). *The state of world fisheries and aquaculture 2008*. Rome: Food and Agriculture Organization of the United Nations (FAO).

Steffen, W., Sanderson, A., Tyson, P. D., Jager, J., Matson, P. M., Moore, B., III, et al. (2004). *Global change and the earth system: A planet under pressure*. New York: Springer.

Tacon, A. G. J., & Metian, M. (2009). Fishing for feed or fishing for food: Increasing global competition for small pelagic forage fish. *Ambio, 38*, 294–302.

Ternström, I. (2005). HIV/AIDS – The true tragedy of the commons? Exploring the effects of HIV/AIDS on management and use of local natural resources. Discussion paper 200. Stockholm: The Beijer Institute of Ecological Economics.

Weber, C. L., & Matthews, H. S. (2008). Food-miles and the relative climate impacts of food choices in the United States. *Environmental Science and Technology, 42*, 3508–3513.

WHO. (2009). Global strategy on diet, physical activity and health – Obesity and overweight. Report, World Health Organization. Available at http://www.who.int/dietphysicalactivity/publications/facts/obesity/en (Retrieved October 23, 2009).

Worm, B., Barbier, E. B., Beaumont, N., Duffy, J. E., Folke, C., Halpern, B. S., et al. (2006). Impacts of biodiversity loss on ocean ecosystem services. *Science, 314*, 787–790.

Chapter 19
Askö 1998: The Value of Nature and the Nature of Value*

Gretchen C. Daily, Tore Söderqvist, Sara Aniyar, Kenneth Arrow, Partha Dasgupta, Paul R. Ehrlich, Carl Folke, AnnMari Jansson, Bengt-Owe Jansson, Nils Kautsky, Simon Levin, Jane Lubchenco, Karl-Göran Mäler, David Simpson, David Starrett, David Tilman, and Brian Walker

The world's ecosystems are capital assets. If properly managed, they yield a flow of vital services, including the production of goods (such as seafood and timber), life support processes (such as pollination and water purification), and life-fulfilling conditions (such as beauty and serenity). Moreover, ecosystems have *value* in terms of the conservation of options (such as genetic diversity for future use) (Daily 1997, 1999). Unfortunately, relative to other forms of capital, ecosystems are poorly understood, scarcely monitored, and (in many cases) undergoing rapid degradation and depletion. Often the importance of ecosystem services is widely appreciated only upon their loss.

This is beginning to change, most notably in Australia and Costa Rica. An Australian firm called Earth Sanctuaries, Ltd., was listed on the Australian Stock Exchange in May, making it the world's first conservation company to go public. The US$25-million firm buys up land (90,000 ha so far) and restores native vegetation and wildlife, earning income from tourism, consulting, and wildlife sales.[1] The firm lobbied for and won a change in accounting law so as to include its rare native animals as assets. Meanwhile, the Sydney Futures Exchange is positioning itself to be a global leader in the trading of ecosystem services, from carbon sequestration (the withholding of carbon, a greenhouse gas constituent, from the atmosphere by plants and soils) to "new environmental products," such as credits for clean water and biodiversity. The CEO of State Forests of

---

* This article was published in *Science*, Vol. 289. Daily et al., The value of nature and the nature of value, 395–396, Copyright AAAS (2000).

[1]Earth Sanctuaries, Ltd., Stirling, South Australia. See www.esl.com.au.

G.C. Daily
Department of Biological Sciences, Stanford University, Stanford, CA 994305, USA

T. Söderqvist, S. Aniyar, K. Arrow, P. Dasgupta, P.R. Ehrlich, C. Folke,
A. Jansson, B.-O. Jansson, N. Kautsky, S. Levin, J. Lubchenco, K.-G. Mäler,
D. Simpson, D. Starrett, D. Tilman, and B. Walker
The Beijer Institute, the Royal Swedish Academy of Sciences, Box 50005,
S-104 05 Stockholm, Sweden

DOI 10.1007/978-90-481-9476-6_19, © Springer Science+Business Media B.V. 2011

New South Wales is promoting a vision of foresters marketing a wide array of ecosystem services, with timber as a "by-product" (Hosking et al. 2000). The Common-wealth Scientific and Industrial Research Organization and The Myer Foundation have just launched the most advanced assessment of ecosystem assets in the world (Cork and Shelton 2000).

Since 1997, the government of Costa Rica has been paying landowners for several ecosystem services: carbon sequestration and protection of water-sheds, biodiversity, and scenic beauty. The payments, about US$50/ha-year, are financed in part by a tax on fossil fuels and are resulting in significant forest conservation and restoration (Castro et al. 1998). Costa Rica has also sold carbon sequestration credits to several European nations. These and other promising government initiatives are supported by scientific expertise and growing industry participation (Gámez Lobo 1999).

Worldwide, ecosystems are being protected or restored to control floods, to filter water, to enhance soil fertility, to stabilize climate, to offer human enjoyment, and even to recycle orange peels (Heal in press; Echavarria and Lochman 1999; Janzen 1999; Reid in press; Masood and Garwin 1998). Such efforts are being rewarded with innovative financial mechanisms, whose scope and variety are expected to grow (see table).

These developments all involve putting a price tag on nature, an act seen by many as risky at best (Ludwig 2000; Gatto and De Leo 2000). To be sure, individuals and societies already assess the value of nature implicitly in their collective decision-making, too often treating ecosystem services as "free." Until recently, this was generally safe to do: relatively speaking, ecosystem capital was abundant, and the impacts of economic activity were minimal. Ecosystem capital is becoming ever scarcer (Vitousek et al. 1997), however, so that it is now critical to understand both how to value ecosystems and the limitations of such valuations.

Ecosystem goods and services	
Commodity	Share of farm business (%)
Wheat	40
Wool	15
Water filtration	15
Timber	10
Carbon sequestration	7.5
Salinity control	7.5
Biodiversity	5

Value of ecosystem goods and services. A hypothetical Australian farm business in 20 years (Binning and Smith 2000). In this model, traditional agricultural commodities account for 55% of revenues, as opposed to 100% today. Other income derives from a mature market for ecosystem goods and services.

19.1 Ecological Basis for Valuation

To establish sound policy, the "production functions" describing how ecosystems generate services need to be characterized, and the interactions among these functions quantified. To begin, a cataloging of the sources and consumers of ecosystem services is needed. For any given location, this would document service flows occurring locally (such as pest control, serenity), across regions (such as timber export and flood control), and globally (such as climate stabilization).

The production functions would also reveal critical points and interdependencies in the supply of services and in the time scales over which services are amenable to repair. Yet these are poorly known now and are likely to remain elusive. Ecosystems typically respond nonlinearly to perturbation. For example, gradual increases in salinity for decades went unnoticed by farmers in Australia but have now reached crisis levels. Replanting native vegetation reduces soil salinity (a benefit) but also reduces river flow (a cost). Furthermore, ecosystems are idiosyncratic; what holds true in one region may not apply well elsewhere. Soil salinity appears controllable with ecosystem approaches in eastern parts of Australia, for example, but in Western Australia, the threshold is higher, and there is little hope for reversal without enormous investment. Putting theory into practice will therefore require locally based information.

19.2 Principles of Valuation

There are three fundamental steps of decision-making. In this context, all require integration of ecological and economic understanding. The first step, identification of possible alternatives, is probably the most important but also the most underrated. Often the identification of alternatives is guided by narrow conventions: if a city is expanding its water treatment system, engineers may evaluate different physical treatment plants, ignoring ecosystem approaches (watershed or wetland management).

The second step requires that all impacts be identified and measured for each alternative: everything from immediate needs for labor, capital, and other inputs to long-term biophysical and social impacts. Rarely does sufficient knowledge exist to make precise estimates, but it is important to try to quantify uncertainties and the risks of proceeding.

The final step, valuation, translates the consequences of maintaining the status quo and opting for each alternative into comparable units of impact on human well-being, now and in the future. These impacts are defined in terms of the resources that people would be willing to forgo to get the goods, services, or other outcomes associated with a particular alternative. The common measuring unit is typically monetary.[2]

---

[2] There is a good case for using multiple numeraires in combination (Schneider et al. in press).

Embedded in this process are several general principles: (i) Public policy decisions involve making incremental, not revolutionary, changes to the status quo. Calculating the total value of ecosystem services, by contrast, is not very helpful. (ii) In a democratic society, values used in social decision-making ought to be derived from those held by its individual citizens and ought not to be imposed by the state. (iii) We should infer people's values as they are revealed by actual decisions whenever possible.

19.3 Scope and Limitations of Valuation

In practice, valuation of ecosystem assets involves some of the oldest problems in economics: revealing and aggregating preferences, and addressing uncertainty. There are drawbacks associated with most ways of inferring value. Market prices often do not reflect the full social costs of production (Naylor et al. 1998, in press); moreover, most services are not presently traded on markets. Methods of indirect revealed preference (for example, valuing clean air by comparing land rents in clean versus polluted areas) are not relevant to setting a value on the existence of certain assets (such as the satisfaction derived from contemplating the existence of a tropical rainforest). Approaches based on avoidance of costs (for example, valuing natural water purification at the cost of its technological alternative, a filtration plant for instance) provide only partial, lower bound indications of value, especially for services without adequate substitute (such as global climate regulation). Contingent valuation surveys (that try to elicit how individuals value hypothetical incremental changes) are improving but still notoriously unreliable, especially when applied to issues with which the public is unfamiliar.

Reliance on individual preferences to construct social values, although defensible on ethical grounds, has serious pitfalls. Preferences depend on institutional context — how much individuals know about the environment, for instance (Costanza 2000). The outcome of economic valuation is in this respect not more informed than the people whose values are being assessed.

Even if we were able to measure individual values accurately, we still must determine how to aggregate these into a social value. Ultimately, the weights used involve a value judgment; there is no "correct" answer. Treating all people equally is appealing in principle but by no means universally accepted.

Measurement of incremental values works best when the increments are small, so that a change in one service will have minimal feedbacks through the rest of the system. Values of various increments can then be estimated separately and simply added. Unfortunately, this condition is difficult to meet for ecosystem services, where the underlying systems tend to be highly interdependent, and seemingly small changes in one place cause large impacts on the overall system.[3] The level of uncertainty in our understanding of ecological processes suggests that it would be

[3] For example, small changes in the phosphate loading in the lake can cause it to flip from a clean state to one of high turbidity (Scheffer 1998).

prudent to avoid courses of action that involve possibly dramatic and irreversible consequences and, instead, to wait for better information.

Another key problem is the relative weight put on current versus future costs and benefits. The choice of "discount rate" is very important where a long time frame is involved; sufficiently high discounting can be used to justify policies that exploit resources now at the expense of substantial environmental costs later. Individuals tend to discount their own futures, whereas "equal treatment" would have future generations treated the same as current ones. Some social discounting is consistent with such equity if future generations will be better off than current ones, a situation that may not continue to prevail (Arrow et al. 1996).

19.4 The State of the Art

Valuation is a way of organizing information to help guide decisions but is not a solution or end in itself. It is one tool in the much larger politics of decision-making. Wielded together with financial instruments and institutional arrangements that allow individuals to capture the value of ecosystem assets, however, the process of valuation can lead to profoundly favorable effects (Berkes and Folke 1998; Heal 2000).

The rapid institutional change presently under way is inspiring for several reasons. It shows that the most important decisions to get right are those where benefits greatly outweigh costs or vice versa, and in such cases, complete accuracy is unnecessary. For example, by constructing crude lower bound estimates for the value of natural water purification services, municipalities worldwide are determining that preserving or restoring natural services is often preferable to constructing a water filtration plant (Heal in press; Echavarria and Lochman 1999; Janzen 1999; Reid in press; Masood and Garwin 1998; Chichilnisky and Heal 1998; National Research Council 2000). The new initiatives also account for the interdependence of services; in Australia and Costa Rica, for instance, multiple services are being bundled to achieve the desired relative increases in supply via changes in land use. With luck, the protection of well-known or highly valued services (such as salinity control and carbon sequestration) will suffice, for now, in preserving those that are poorly known (such as pollination) (Kremen et al. 2000). Finally, the initiatives are generating demand for, and spurring the development of, integrated ecological–economic–social approaches to managing ecosystem assets, and the potential for such approaches is tremendous.

Acknowledgements We benefited from discussions with C. Binning, J. Eagle, M. Freeman, L. Gámez, and J. Salzman. The Beijer Institute provided the impetus and forum for this work.

References

Arrow, K. et al. (1995). In Bruce, J., Lee, H., & Haites, E. (Eds.). *Climate Change 1995* (pp. 124–144). Cambridge, UK: Intergovernmental Panel on Climate Change and Cambridge University Press.

Berkes, F., & Folke, C. (Eds.). (1998). *Linking social and ecological systems*. Cambridge: Cambridge University Press.
Binning, C., & Smith, R. (2000). Presentations in Hosking et al. (2000).
Castro, R., Tattenbach, F., Gámez, L., & Olson, N. (1998). *The Costa Rican experience with market instruments to mitigate climate change and conserve biodiversity*. San José, Costa Rica: Fundecor and MINAE.
Chichilnisky, G., & Heal, G. (1998). Economic returns from the biosphere. *Nature, 391*, 629–630.
Cork, S., & Shelton, D. (2000). *Sustainable environmental solutions for industry and government*. Proceedings of the 3rd Queensland Environmental Conference, 25 to 26 May 2000, Brisbane, Australia, in press; www.dwe.csiro.au/ecoservices/
Costanza, R. (2000). Social goals and the valuation of ecosystem services. *Ecosystems, 3*, 4–10.
Daily, G. C. (Ed.). (1997). *Nature's services*. Washington, DC: Island Press.
Daily, G. C. (1999). Developing a scientific basis for managing Earth's life support systems. *Conservation Ecology, 3(2)*, 14 (1999). [online] Available at www.consecol.org/vol3/iss2/art14.
Echavarria, M., & Lochman, L. (1999). *Policy mechanisms for watershed conservation*. Arlington, VA: Nature Conservancy.
Gámez Lobo, R. (1999). *De Biodiversidad, Gentes y Utopías*. Santo Domingo, Costa Rica: INBio.
Gatto, M., & De Leo, G. (2000). Pricing biodiversity and ecosystem services: The never-ending story. *BioScience, 50*, 347–355.
Heal, G., in press. *Nature and the marketplace*. Washington DC: Island Press.
Heal, G. (2000). Valuing ecosystem services. *Ecosystems, 3*, 24–30.
Hosking, L., Smith, R., & Brand, D. (2000). Presentations at "Forests in the Green Economy," A conference cosponsored by the Sydney Futures Exchange; Forests Trends; and State Forests of New South Wales, Australia, 10 to 12 April 2000, Sydney, Australia. See also www.forest.nsw.gov.au.
Janzen, D. (1999). Gardenification of tropical conserved wildlands: Multitask, multicropping, and multiusers. *Proceedings of the National Academy of Sciences USA, 96*, 5987–5994.
Kremen, C., et al. (2000). Economic incentives for rain forest conservation across scales. *Science, 288*, 1828–1832.
Ludwig, D. (2000). Limitations of economic valuation of ecosystems. *Ecosystems, 3*, 31–35.
Masood, E., & Garwin, L. (1998). Costing the Earth: When ecology meets economics. *Nature, 395*, 426–427.
National Research Council. (2000). *Watershed management for potable water supply*. Washington, DC: National Academy Press.
Naylor, R., et al., in press. Effect of aquaculture on world fish supplies. *Nature*.
Naylor, R., et al. (1998). Nature's subsidies to shrimp and salmon farming. *Science, 282*, 883–884.
Reid, W. (in press). *Managing human-dominated ecosystems*. Washington DC: Island Press.
Scheffer, M. (1998). *The ecology of shallow lakes*. London: Chapman and Hall.
Schneider, S., Kuntz-Duriseti, K., & Azar, C. (in press). Costing non-linearities, surprises and irreversible events. *Pacific and Asian Journal of Energy*.
Vitousek, P. M., Mooney, H. A., Lubchenco, J., & Melillo, J. M. (1997). Human domination of earth's ecosystem. *Science, 277*, 494–499.

Chapter 20
Askö 1998: Commentary by Shahid Naeem

Shahid Naeem

The functioning of an ecosystem is only tangentially related to the services it provides, a fact that stymies worldwide efforts to mainstream ecosystem services into environmental management in the way Daily et al., a collaboration among ecologists and economists, called for nearly a decade ago. Recall that a system is a set of interacting parts whose collective activities define system functioning, but what services it provides depends on human values. For example, a mechanical clock is a system of interacting gears whose function is to tell time, but its services or values as a chronometer, paper weight, or thing of beauty, are ascribed by humans. In the case of an "ecosystem", its set of interacting parts is its community of species whose collective metabolic activities redistribute chemicals among the solid, liquid, and gaseous spaces they occupy. Some common ecosystem functions include autotrophic production (largely by plants and algal species that consume inorganic carbon, such as CO_2) and heterotrophic production (largely microbial and animal species that consume organic carbon in the form of living or dead organisms). Other ecosystem functions include transfers of many elements, including C, H, N, O, P, S, from one place on Earth to another. For example, terrestrial autotrophic biomass annually removes 59.22 Gt of C (where Gt = gigatons or 10^{15} g) from the atmosphere (Haberl et al. 2007) and this carbon is eventually distributed throughout plant, animal, and microbial species, soil, sediments, and water. This biogeochemistry, or biologically modified geochemistry, constitutes *ecosystem functioning*. An ecosystem also alters the physical attributes of its surroundings, such as its albedo (light reflectance), microclimate (e.g., local humidity and temperature), and hydrology (e.g., water retention). These are *ecosystem properties*. Organisms within ecosystems also regulate each other's densities through *biotic regulation*, such as through herbivory, carnivory, parasitism, and facilitative interactions, such as pollination and mycorrhizal fungal associations that benefit plants. What services ecosystems provide, what properties and biotic interactions we

S. Naeem
Department of Ecology, Evolution and Environmental Biology, Columbia University, New York, NY 10027, USA
e-mail: sn2121@columbia.edu

T. Söderqvist et al. (eds.), *Bringing Ecologists and Economists Together*,
DOI 10.1007/978-90-481-9476-6_20, © Springer Science+Business Media B.V. 2011

value, are ascribed by humans. Daily et al. drew attention to the rapidly emerging synthesis of social and natural sciences that redefined ecosystems. Rather than biodiversity driving biogeochemical flows, altering the physical environment, and regulating densities of species, ecosystems were recast as capital assets that yield flows of services such as the production of goods, life support, and the generation of life-fulfilling conditions. This morphing of the ecosystem concept filled an urgent need to bring nature into our environmental deliberations, especially environmental decision making and management, but its core construct is untenable; ecosystem service flows do not map sensibly onto ecosystem functions, ecosystem properties, or biotic regulation. Consider, for example, that Daily et al. considered ecosystem services, citing the Australian and Costa Rican systems, as biodiversity itself, a practice that is not uncommon in payment for ecosystem services. Yet species are parts of the system (like gears in a clock), not services provided by the system.

For ecologists the utility of Daily et al. is limited because of its lack of ecological coherency, though it has been influential as an idea. Cited by over one hundred publications (ISI Web of Science), there is little doubt that Daily et al., as a science policy piece published in the high profile journal *Science*, has had its impact, but ecology has been conceptually inspired rather than scientifically influenced. To be sure, ecosystem services have become mainstream in ecological thinking (Kremen and Ostfeld 2005; Kremen et al. 2007) and ecosystem services were the centerpiece of the conceptual framework of the Millennium Ecosystem Assessment (MEA 2003) in which most of the world's leading ecologists participated. There has been, however, little in the way of new ecological research pertaining to ecosystem services. Most papers citing Daily et al. are overview, concept, policy, or economic papers; only a handful (8 out of 103) have been ecological research papers. Examples are valuing oxygen and greenhouse gas regulation as services provided by Chinese rice paddies (Xiao et al. 2005), valuing birds for their reduction of herbivory on oil palm in plantations (Koh 2008), evaluating the importance of forests for the production of hydroelectricity (Guo et al. 2007), and the value of pollinators in crop production (Kremen et al. 2007). These are important studies, but they cobble together ecology and economics rather than successfully treat ecosystems as capital assets and convert ecosystem functions, properties, and biotic regulations into ecosystem services. These studies achieve the goals of valuation that Daily et al. strongly emphasized, but they do so by linking known market values for goods and services with ecosystem functions (e.g., oxygen and food production), properties (e.g., the physical property of forest controls over siltation) and biotic regulation (e.g., pollination and the control of herbivory). These studies, though novel, did not stray too far from traditional natural resource economics, like valuing biocontrol and pollination in agriculture using market values for crops.

The lack of traction of ecosystem services in ecological research stems from its incongruence with current ecological understanding. The conceptualization of ecosystem services as developed in the central framework of the MEA, which sees increasing adoption across most sectors, is that plants, animals, and microorganisms (i.e., biodiversity) generate ecosystem functions, and those functions that benefit humans constitute ecosystem services. Human well-being resides at the end of this causal chain. That is, biodiversity → ecosystem functioning → ecosystem services → human well-being. Ecosystem services are divided into supporting services (e.g., elemental flows) that ultimately sustain life on Earth, and provisioning (e.g., timber and food), regulating (e.g., vegetation stabilization of soil), and cultural (e.g., inspirational or recreational values of nature) services. This causal chain, however, is illogical. Consider Fig. 20.1, which attempts to simultaneously portray ecological and service flows in more detailed fashion than the simple causal chain. In contrast to the MEA framework, note that that supporting services concern linkages between the geochemical and biological components of ecosystems and not people. In contrast, provisioning, regulating, and cultural services concern ecological flows from biodiversity to people. Note also that people influence supporting services by their influence on geochemical processes (e.g., fertilizer addition) and other services by their influences over biodiversity (e.g., substituting domesticated plants for native plants).

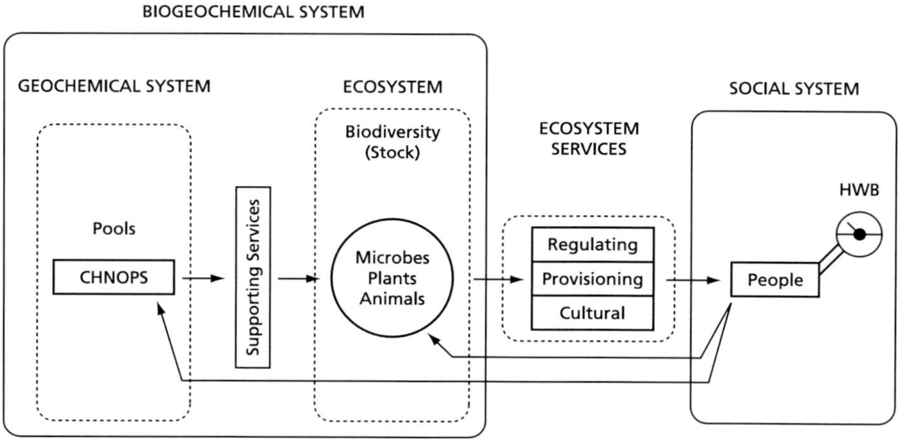

Fig. 20.1 Ecological inconsistencies in the conceptualization of ecosystem services. Ecosystem services, as defined by Daily et al., are service flows from ecosystems, classified by the Millennium Ecosystem Assessment as supporting, regulating, provisioning, and cultural. Ecologically, ecosystems are aggregates of plant, animal, and microbial species whose metabolic activities influence the flows of elements (such as C, H, N, O, P, and S) in the geochemical systems of Earth. Humanity does not influence service flows directly, but indirectly. They influence supporting services by the way they influence geochemistry (e.g., by industrial injection of greenhouse gasses into the atmosphere) and the remaining services by the way they influence biodiversity (e.g., substituting domestic species for natural species in farms and aquaculture or over harvesting fish and wildlife). Humans reside in a social system in which the state of human well-being (HWB, here illustrated as a meter connected to humanity) influences elemental pools and biodiversity through a complex set of decisions. The standard assumption that biodiversity determines ecosystem functioning which translates directly into ecosystem services which, in turn, influences human well-being, is inconsistent with the more complex pattern of linkages that structure coupled social–natural systems. For example, supporting services are distinct from other services and people are part of biodiversity and reside in ecosystems as well as social systems – even this figure fails to capture the complexities inherent in ecosystem services

While these additional complexities yield a more realistic construct for ecosystem services, they also point to two fundamental problems in any treatment of ecosystem services. First, ecosystems do not exist without biodiversity, and biodiversity does not exist outside ecosystems. Second, it is not logical to separate humans from biodiversity as we too are biological entities and are directly linked to supporting services (i.e., we need oxygen as much as any aerobic organism). It is illogical to separate biodiversity, ecosystems, and humans, thus any construct of ecosystem services that is based on such separations is not tenable.

These problems are not insurmountable, and at a time when there is a global call to abandon traditional development and strive for building a sustainable world, ecosystem services will play a key role in achieving this goal. Much of the history of human development has been defined as the building up of human, manufactured, and financial capital at the expense of natural capital (Chee 2004), a trend that is the root of the most pressing environmental challenges humanity faces, as recognized over 2 decades ago in the United Nations' Brundtland Commission report (WCED 1987).

The concept of ecosystem services can play a pivotal role in reshaping economic development because it provides the means by which service flows from natural assets can be interwoven into the environmental decisions we make. Daily et al. argued that effective environmental decision making hinged on identifying all options, knowing the costs and benefits of pursuing different options, and constructing valuations that would clarify the consequences of inaction as well as the relative benefits of pursuing different options. Using ecosystem services, no longer would a landowner's decision about use of her forest be based solely on market values for timber, the traditional route for deforestation. Instead, payments for ecosystem services, such as carbon sequestration, watershed protection, biodiversity, and scenic beauty, made by governments, other institutions, or financial instruments, would provide an array of options that truly reflected the broader value of her forest. Such valuation will be only as good as its ecology, but that ecology has yet to be developed.

The message of Daily et al. is important, however, even though it currently lacks congruity with modern ecological understanding. Ecologists and economists need to work together to fix this problem. Our success as a species and our future well-being has been intimately tied to a 3.5 billion year old system governed by ecological and evolutionary processes. We have been part of a global ecosystem collective – one in which grasslands, forests, lakes, rivers, estuaries, and other ecosystems exchange energy and materials with each other. This collective, our Biosphere, is enormous, estimated to be roughly 10^3 Gt in mass, roughly divided equally between microbial and multicellular life (Whitman et al. 1998). It is also complex, comprising some 10–100 million species (Groombridge and Jenkins 2000). The antiquity and complexity of this system is incomprehensible. About six million years ago, humans became distinct among other primates and sometime later, they began deliberately manipulating biodiversity and ecosystem functioning in ways that improved their well-being. Only 11,000 years ago, however, when agriculture and animal domestication began, we began dramatically altering biodiversity and ecosystem function in terrestrial ecosystems (Miller 2008). Now we have begun domesticating marine species (Duarte et al. 2007) and are relentlessly harvesting marine resources (Worm et al. 2006).

Today, humans dominate all ecosystems (Vitousek et al. 1997) and are rapidly domesticating the entire world (Kareiva et al. 2007). We consume a quarter of terrestrial production (Haberl et al. 2007) and we are driving biodiversity to extremely low levels. Taking this separate path, one in which we turn ecosystem services to our purposes rather than being ruled by ecological and evolutionary processes, has been the source of our success, but also the source of our most pressing problems such as emerging diseases, climate change, and world food and energy shortages. Daily et al. correctly point to ecosystem services as a critical component of effective environmental management, particularly with respect to valuation, but the ecological foundation of ecosystem services, which is currently flawed, needs further development. This is something ecologists and economists, working in concert, can and should achieve in the next few years.

References

Chee, Y. E. (2004). An ecological perspective on the valuation of ecosystem services. *Biological Conservation, 120*, 549–565.

Duarte, C. M., Marba, N., & Holmer, M. (2007). Rapid domestication of marine species. *Science, 316*, 382–383.

Groombridge, B., & Jenkins, M. D. (2000). *Global biodiversity: Earth's living resources in the 21st century*. Combridge, UK: World Conservation Press.

Guo, Z., Li, Y., Xiao, X., Zhang, L., & Gan, Y. (2007). Hydroelectricity production and forest conservation in watersheds. *Ecological Applications, 17*, 1557–1562.

Haberl, H., Erb, K. H., Krausmann, F., Gaube, V., Bondeau, A., Plutzar, C., et al. (2007). Quantifying and mapping the human appropriation of net primary production in earth's terrestrial ecosystems. *Proceedings of the National Academy of Sciences, 104*, 12942–12947.

Kareiva, P., Watts, S., McDonald, R., & Boucher, T. (2007). Domesticated nature: Shaping landscapes and ecosystems for human welfare. *Science, 316*, 1866–1869.

Koh, L. P. (2008). Birds defend oil palms from herbivorous insects. *Ecological Applications, 18*, 821–825.

Kremen, C., & Ostfeld, R. S. (2005). A call to ecologists: Measuring, analyzing, and managing ecosystem services. *Frontiers in Ecology and the Environment, 3*, 540–548.

Kremen, C., Williams, N. M., Aizen, M. A., Gemmill-Herren, B., LeBuhn, G., Minckley, R., et al. (2007). Pollination and other ecosystem services produced by mobile organisms: A conceptual framework for the effects of land-use change. *Ecology Letters, 10*, 299–314.

MEA. (2003). *Ecosystems and human well-being: A framework for assessment. Millennium Ecosystem Assessment*. Washington, DC: Island Press.

Miller, F. P. (2008). After 10,000 years of agriculture, whither agronomy? *Agronomy Journal, 100*, 22–34.

Vitousek, P. M., Mooney, H. A., Lubchenco, J., & Melillo, J. M. (1997). Human domination of earth's ecosystem. *Science, 277*, 494–499.

WCED. (1987). *Our common future. Report of the world commission on environment and development*. Oxford: Oxford University Press.

Whitman, W. B., Coleman, D. C., & Wiebe, W. J. (1998). Prokaryotes: The unseen majority. *PNAS, 95*, 6578–6583.

Worm, B., Barbier, E. B., Beaumont, N., Duffy, J. E., Folke, C., Halpern, B. S., et al. (2006). Impacts of biodiversity loss on ocean ecosystem services. *Science, 314*, 787–790.

Xiao, Y., Xie, G., Lu, C., Ding, X., & Lu, Y. (2005). The value of gas exchange as a service by rice paddies in suburban Shanghai, PR China. *Agriculture, Ecosystems and Environment, 109*, 273–283.

Chapter 21
Askö 1998: Commentary by Jeffrey R. Vincent

Jeffrey R. Vincent

The article by Daily et al. drew attention to the multi-faceted value of ecosystems to human beings, and it highlighted some of earliest examples of payments for ecosystem services (PES). These innovative financial mechanisms convert latent economic values into actual monetary flows, thereby strengthening incentives for conservation. Daily et al. noted that PES "involve putting a price tag on nature," which in turn requires knowing which services ecosystems provide and understanding how human activities affect the quantity and quality of those services. Although they cautioned that "Valuation…is not a solution or end in itself," they concluded that "Wielded together with financial instruments and institutional arrangements that allow individuals to capture the value of ecosystem assets…the process of valuation can lead to profoundly favorable effects."

Interest in PES has grown since the article was published. Numerous publications describe more recent examples (e.g., Pagiola et al. 2002), and an online clearinghouse for information on them now exists (the Katoomba Group's "Ecosystem Marketplace", http://www.ecosystemmarketplace.com). At the same time, the importance of valuation for conservation has come under attack. The best-known statement of this skeptical view comes from Heal (2000, p. 126), who wrote that

> Incentives are critical for conservation. Valuation is not necessary for establishing the correct incentives, although correct incentives may provide a mechanism for valuation.

According to this view, government agencies and conservation organizations need to get on with the business of creating PES systems, which through their operation will reveal societal values for nature. Although valuation studies may have academic value, they have little conservation value.

This view understates the potential importance of valuation studies for designing PES systems. Although a valuation study may not need to be done before every PES system is implemented, valuation studies can help raise the success rate

J.R. Vincent
Nicholas School of the Environment and Sanford School of Public Policy, Duke University, USA
e-mail: jeff.vincent@duke.edu

T. Söderqvist et al. (eds.), *Bringing Ecologists and Economists Together*,
DOI 10.1007/978-90-481-9476-6_21,

for PES systems by providing designers with information on the demand for ecosystem services. By this I mean not just a single number – how much is an ecosystem worth? – but rather the relationship between households' willingness to pay and the characteristics of ecosystem services, their own socioeconomic characteristics, and institutional factors. Nasi et al. (2002, p. 2) usefully restate the quote by Heal as follows:

> Incentives are critical for conservation: valuation is not sufficient for establishing the correct incentives, although it may be a prominent tool to reveal the relevant incentive structures.

Valuation skeptics have perhaps been unduly influenced by the dominant features of many ecosystem valuation studies, which instead of yielding demand functions have instead yielded point estimates of the "total economic value" of particular ecosystems. Although these estimates can, when plugged into benefit-cost analyses that compare conservation to conversion (e.g., mangroves vs. shrimp farms, rainforests vs. ranches, floodplains vs. subdivisions), help direct the creation of PES systems toward the economically most important opportunities, the "wielding together" of valuation and financing can, and should, involve much more. Valuation studies can identify households' preferences for different types and levels of ecosystem services. Most conservation decisions are not about the binary choice of whether to protect or to convert, but rather about the selection of the best ecosystem management option out of a set that includes many. (See Holmes and Boyle 2003, for a forestry example linked to valuation.) This selection affects the scale and scope of revenue flows from PES systems. Better knowledge about the set of options and household preferences for them can be used in designing PES.

The role of valuation in generating information necessary for selecting among different service options is seen most clearly in a context outside of nature conservation. I refer to the 2 decades of research by Dale Whittington and colleagues on household willingness to pay for improved municipal services, specifically clean water and sanitation, in developing country cities (e.g., Whittington et al. 1990). There are multiple options for improving water and sanitation services to households, and different households are willing to pay different amounts for these options. The valuation studies by Whittington and colleagues have generated practical information that municipalities and aid agencies have used to determine which services to provide, what levels to provide them at, how to structure payments (e.g., different rate structures for piped water), and whether or not the systems will be self-financing.

Such demand-oriented valuation studies are less common in the case of nature conservation, but some examples exist. Most pertain to entrance fees for parks. A good example is a study on tourists' willingness to pay to view wildlife in Kenya, by Navrud and Mungatana (1994). The Kenya Wildlife Service used the study's findings to justify a quadrupling of entrance fees. Shultz et al. (2000) conducted a similar study in Costa Rica. Interestingly, their study also analyzed whether proposed improvements in park infrastructure and services – in effect, the delivery mechanism for the ecosystem services that lure visitors to parks – could be financed through increases in entrance fees, which of course turns on the elasticity of demand. A study by Holmes et al. (1998) provides another example of a valuation study that generated information on demand for specific park attributes, in this case in Brazil.

In all these studies, the type of payment mechanism, a park entrance fee, was given. Valuation studies can also be used to determine which payment mechanism is the best one for particular ecosystem services. It is well understood by public finance economists that payment mechanisms affect the amount of revenue generated. This is one of the fundamental reasons governments employ general tax revenues to pay for some public services, earmarked taxes for others, and user fees for yet others. In some developing countries, the public's lack of trust in the government might lead them to be willing to pay very little if revenues from a PES system are collected and managed by a government agency. On the other hand, they might be willing to pay much more if an NGO is responsible for these functions. Valuation studies can shed light on the effect of different payment mechanism options on the ability of PES systems to generate financial flows needed for conservation.

Valuation studies can thus assist in the design of PES systems, parallel to the way that marketing studies help private companies decide on the features of new products. This design role is important, because experience with PES systems since the Daily et al. article was published reveals a mixed record. The most famous example of a PES at that time was, and still is, New York City's investment in improved watershed management in the Catskills to reduce nonpoint source pollution in lieu of building a filtration plant to treat surface waters. According to Sagoff (2002), claims that New York City floated an environmental bond issue to finance the watershed program and bought some 100,000 acres of land in the Catskills are no more than urban legends. There was no bond issue, and the city had purchased less than 20,000 acres by February 2002.

In summary, PES systems won't work unless beneficiaries are willing to make payments for the services they receive. Valuation studies can help predict whether beneficiaries are willing to pay and, more important, how their willingness to pay depends on the characteristics of the delivery and payment mechanisms that are key components of PES systems.

References

Heal, G. M. (2000). *Nature and the marketplace*. Washington, DC: Island Press.

Holmes, T. P., Alger, K., Zinkhan, C., & Mercer, D. E. (1998). The effect of response time on conjoint analysis estimates of rainforest protection values. *Journal of Forestry Economics, 4*(1), 7–28.

Holmes, T. P., & Boyle, K. J. (2003). Stated preference methods for valuation of forest attributes. In E. O. Sills & K. L. Abt (Eds.), *Forests in a market economy*. Dordrecht, the Netherlands: Kluwer Academic.

Nasi, R., Wunder, S., & Campos, J. J. (2002). Forest ecosystem services: Can they pay our way out of deforestation? Executive version of a discussion paper prepared for the GEF Forestry Roundtable held in conjunction with the UNFF II in New York on March 11, 2002. Center for International Forestry Research, Jakarta, Indonesia. http://www.unep.org/dec/docs/Forest_Ecosystem_Service-Executive_Version.pdf (accessed June 16, 2009).

Navrud, S., & Mungatana, E. D. (1994). Environmental valuation in developing countries: The recreational value of wildlife viewing. *Ecological Economics, 11*, 135–151.

Pagiola, S., Bishop, J., & Landell-Mills, N. (Eds.). (2002). *Selling forest environmental services.* London: Earthscan.

Sagoff, M. (2002). On the value of natural ecosystems: The catskills parable. *Politics and the Life Sciences, 21*(1), 19–25.

Shultz, S., Pinazzo, J., & Cifuentes, M. (2000). Opportunities and limitations of contingent valuation surveys to determine national park entrance fees: Evidence from Costa Rica. *Environment and Development Economics, 3*(1), 131–149.

Whittington, D., Briscoe, J., Mu, X., & Barron, W. (1990). Estimating the willingness to pay for water services in developing countries: A case study of the use of contingent valuation surveys in Southern Haiti. *Economic Development and Cultural Change, 38*(2), 293–311.

Chapter 22
Askö in Washington 1999: Managing Ecosystem Resources*

Kenneth Arrow, Gretchen Daily, Partha Dasgupta, Simon Levin, Karl-Göran Mäler, Eric Maskin, David Starrett, Thomas Sterner, and Thomas Tietenberg

We explore some of the special problems faced in the management of environmental resources, paying particular attention to valuation of ecosystem services, externalities, uncertainty, and nonlinearities characteristic of complex adaptive, highly interconnected systems. Through consideration of case studies drawn from the management of lake and mangrove ecosystems, we develop a theoretical perspective in which we analyze the challenges, suggest approaches to their resolution, and endeavor to derive principles that may guide management more generally.

*This article was published in *Environmental Science and Technology*, Vol. 34, Arrow et al., Managing ecosystem resources, 1401–1406, Copyright American Chemical Society (2000).

K. Arrow, G. Daily, and D. Starrett
Department of Economics (Arrow and Starrett) and Department of Biological Sciences (Daily), Stanford University, Stanford, CA 94305, USA

P. Dasgupta
Faculty of Economics and Politics, Cambridge University, Cambridge CB3 9DD, UK

S. Levin
Department of Ecology and Evolutionary Biology, Princeton University, Princeton, NJ 08544–1002, USA

E. Maskin
Department of Economics, Harvard University, Littauer 308, Cambridge, MA 02138, USA

T. Sterner
Resources for the Future, 1616 P Street, Washington, DC 20036, USA

T. Tietenberg
Department of Economics, Colby College, Waterville, ME 04901, USA

K.-G. Mäler
Beijer Institute of Ecological Economics, the Royal Swedish Academy of Sciences, Box 50005, SE-10405 Stockholm, Sweden

T. Söderqvist et al. (eds.), *Bringing Ecologists and Economists Together*,
DOI 10.1007/978-90-481-9476-6_22, © Springer Science+Business Media B.V. 2011

22.1 Introduction

Humans are part of nature and must utilize the bounty it provides in order to survive. However, the choices we make regarding how to utilize natural systems have fundamental implications for their maintenance and ultimately for the sustainability of the services they provide humans. We rely on natural systems directly for food, water, oxygen, fiber, fuel, and pharmaceuticals and indirectly for pollination, for the stabilization of climates and coasts, and for an uncountable list of other essential aspects of our quality of life (Daily 1997). We deliberately modify some systems (as in agriculture) and extract resources from others, in all cases affecting a range of services these systems do and might provide.

The exploitation of natural systems, including efforts to modify and manage them, forces us to confront the tradeoffs between real and potential services and the effects upon their resiliency.

Although it is appropriate and relatively straightforward to construct lists of the most important of such services, their valuation – an essential step for making management decisions – poses daunting challenges. Despite some efforts in that direction (Costanza et al. 1997), it makes no sense to speak of the total value of ecosystem services on the planet. What does make sense, however, is to attempt to estimate the marginal costs that would be associated with having to replace the services currently provided by a piece of nature. The challenge, nonetheless, is far from easy. Conflicting value systems, intergenerational and intragenerational equity, and the basic principle of maintaining the integrity and resiliency of ecosystem functions pose daunting tasks. Our goal in this paper will be to identify some of the problems and to illustrate them with particular examples that exhibit the essential complexities. Our approach represents a theoretical perspective, illuminated by case studies, rather than an effort to deal with all the practicalities of those case studies.

In principle, efficient management of ecosystems involves the same economic principles as does efficient management of fossil fuels and other capital assets. In practice, however, ecosystems possess several features that make good management particularly problematic. They are, first of all, highly nonlinear complex adaptive systems (Levin 1998, 1999), with extensive interconnections among components. Such features lead to the existence of multiple domains of attraction, to elaborate potential path dependency in development, and to the possibility of qualitative shifts in dynamics due to a combination of endogenous and exogenous factors. Periods of drought in grasslands, for example, can lead to patterns of erosion, loss of tree species, and eventual desertification if extended across long enough periods of space and time. Similar major transitions can also occur in aquatic and marine systems due to positive feedbacks, and we will return to some of these examples later. Our conclusions will be that, given the challenge of sustaining services and managing ecosystems in the face of exogenous and endogenous uncertainty, such nonlinearities and unpredictable aspects argue for adaptive management and precautionary principles.

A related feature is that ecosystem development is an idiosyncratic, historically constrained process, making generalization difficult and uncertainty high. Finally, regarding human exploitation, ecological assets in large part represent public goods, and the consequent externalities associated with their use are not usually well accounted for in market mechanisms. In this paper, we will attempt to elucidate these difficulties and suggest ways of mitigating their effects. In so doing, we will draw on two specific examples: mangrove swamps in the tropics and shallow lakes in North America.

22.2 Managing a Mangrove Ecosystem

Intelligent management of any system requires quantification of costs and benefits and evaluation of the tradeoffs involved in different courses of action. The most powerful way to do this bookkeeping and projection is through some sort of model, which requires as a first step the construction of a systematic, quantitative catalog of the sources and consumers of ecosystem services. For a given location, one must know which services are used locally (e.g., pollination, pest control, renewal of soil fertility, serenity); which are used globally (e.g., preservation of the genetic library, climate stabilization), and which are exported to other regions (e.g., seafood, timber, flood control, water purification). For the case of a mangrove swamp, some of these are obvious, but some are hidden. In the category of directly consumed goods, the mangroves harbor and sustain fisheries containing not only species that are endemic to the swamp but also others that migrate and are only occasional users. The mangrove trees themselves are used as wood for fuel and for construction. Salt can also be produced in some circumstances.

The mangroves also provide many indirect services. By providing buffers to the coastline, they offer storm control. By anchoring the soil, they reduce erosion. They recycle nutrients, help in the purification of sewage, and provide habitat for a variety of birds; partly as a consequence, they also can serve as an attraction for tourism. Mangroves also sequester carbon as well as toxic materials. Although this is only a partial list, one can see that the dimension of the output space in such systems can be quite large. In and of itself, that fact is not impossible to overcome – after all, the output space of a typical firm contains a large number of products. However, when it comes to measuring and comparing all these effects, the problems are more serious. The firm knows how to measure its products, and the profit motive provides a natural way of comparing the importance of various items; such ready measures, however, are not easily available for ecosystems.

Despite these remarks, we can formulate a general model for analyzing the impact on well-being from different management strategies. We posit the existence of a social planner who represents society in its preferences for services supplied by the mangrove swamp. The Nobel Prize winning economist Tjalling Koopmans showed in a number of articles that these preferences can be represented as the "social well-being":

$$W = \int_0^\infty U(C(t))e^{-\delta}t\,dt \tag{22.1}$$

where $C(t)$ is a vector of all the relevant consequences from changes in management strategies. The components may be the catch of fish at time t, the amount of fuel-wood collected in the swamp, etc. We are assuming here a sufficient degree of altruism that, for practical purposes, we can consider government policy as seeking to maximize the common ethical principles of humanity. In reality, the results we seek only require Pareto optimality, the discussion in terms of a social planner being solely for convenience in presentation.

All these consequences are aggregated into the utility at time t by the function U. Typically, U is a concave function, reflecting diminishing returns. This implies that the gains from higher than average utilization in some years does not compensate for the losses in less than average years; hence, this implies that, everything else being equal, society is interested in a more equal distribution between different time periods. W is the present value of future flows of utility, discounted at rate $\delta > 0$. Note that this rate is the utility discount rate, which in general is different from the rate at which consumption (or the monetary value of the consequences) is discounted. That the utility discount rate is positive is, according to Koopmans, a mathematical necessity. Otherwise, we would not be able to define preferences for the consequences from now and into the infinite future. Determining the discount rate can be problematical, however, especially given issues of intergenerational equity. The discount rate (which more generally need not be constant as the time horizon expands) is perhaps the most influential parameter governing cost-benefit analysis and management choices, yet its proper choice is extremely difficult.

We can now define the optimal management as a choice of all the control variables that determine the future development of the mangrove swamp, including all the services provided by the swamp, as being a choice that maximizes social welfare. The choice need not be fixed for all time but is a best estimate based on current information. In adaptive approaches, continual updating may be necessary. The social planner thus chooses the use of the mangrove trees, the fishing effort, etc. so as to maximize the present value of the stream of future utilities. To do that, one obviously needs information on the working of the ecological systems and on the utility derived from the services provided by the mangroves. With this in mind, let us now return to the comparison with the multiproduct firm.

There are several reasons why the problem of managing an ecosystem is different in character from that of managing a multiproduct firm. For one thing, the producers in the firm generally have good knowledge of the products that they can (potentially) produce and of the production technology for doing so. By contrast, we are relatively ignorant concerning relationships in ecosystems and are likely to underestimate the list of services they provide. This represents a particular form of uncertainty – what unknown benefits, such as potential pharmaceuticals yet to be identified, lie concealed from our view, awaiting discovery?

A second difficulty with attempting to manage an ecosystem like a multiproduct firm is that, for the firm, the future utilities are given by the forecasted market prices for the goods it is producing; in contrast, for the mangrove, system many of the outputs have a "public" or "common property" feature that makes it difficult to determine the relative value the service conveys or to appropriate that value to the entity doing the managing.

For example, storm protection has this public character. If it is provided, everyone inland automatically gets the benefits. If any particular farm had to contract for such services, it would find that most of the benefits would go to others (free riders); thus, there would be little incentive to participate. Contrast this with wood for fuel, which is similar in character with most of the products we associate with the multiproduct firm. This wood is private, in that the benefits of its use accrue to the person who gets it. Consequently one can contract for it, and the price paid will be a reasonably good measure of the value it has to the person who values it most highly. Fish are a common property resource to the fishermen but take on private goods aspects once caught. Thus, the fish can be contracted for, but the common resource must be managed centrally if we are to avoid overexploitation of the resource. Among the other services identified in the mangrove swamp, soil erosion protection and nutrient recycling have very much the same public character as storm protection. Carbon sequestration is an extreme example where the benefits are not even confined locally but accrue to all of humanity. Indeed, the beneficiaries of conservation actions designed to sequester carbon are in general separated spatially and temporally from the costs of those actions. Determining the values to place on such items, relative say to firewood, is quite difficult but must be done if there is to be any "bottom line" for managers to pursue.

The next step in effective modeling is to understand the dynamic relationships among the various variables of the system. This step is analogous to knowing the production relationships in a multiproduct firm. Some variables of an ecosystem are fixed at a moment of time and can only be changed with the passage of time. These we refer to as the *state* variables of the system. For the mangroves these will include the area of the swamp (A), the biomass of the mangroves (M), the fish stock(s) (Y), and the stock of nutrients in the soil (N). Other variables of the system (flow or control variables) can be changed at a moment of time. In the mangroves, control variables will include the number of person and boat hours devoted to fishing for the various species, the number of fish caught per unit time (h_y), the amount of wood cut for fuel and construction materials (h_m), and the number of boats used for sightseeing.

Once we have identified the relevant variables, we need to characterize the interdependencies among them and, consequently, the ways in which they will evolve over time as functions of choices we make. From a mathematical point of view, the result will typically be a system of differential equations that tell us the rates of change of the state variables at any moment of time as a function of the current state and values of flow variables. In the absence of uncertainty, these equations would determine the transformation of the ecosystem over time.

For the mangroves, we can start to construct the necessary system as follows. To begin, we model the local, near-shore fish population in terms of two functions, the unimpeded growth rate (g_y) and the carrying capacity (C_y). This is a drastic oversimplification, neglecting such complexities as the age structure or size structure of the population; but it is a place to begin. The unimpeded growth rate is in itself a compound function, collapsing the rates of recruitment and mortality into a single variable; furthermore, it will depend on the state variable and extrinsic variables in ways often difficult to quantify. Furthermore, for fish endemic to the mangroves, both of these parameters will depend on the area A of the swamp and stock of nutrients N. Thus, simplistically, the dynamics of the fish population are assumed to change according to a logistic relation of the form

$$\mathrm{d}Y / \mathrm{d}t = g_y(A,N)Y(C_y(A,N)\text{-}Y)\text{-}h_y \tag{22.2}$$

where Y is the size of the fish population and h_y is the harvest rate. Similarly, the mangroves M will have their own growth rate g_m, their own carrying capacity, and their own harvest rate h_m, leading to an equation of the form

$$\mathrm{d}M / \mathrm{d}t = g_m(A,N)M(C_m(A,N)\text{-}M)\text{-}h_m \tag{22.3}$$

for the rate of change of biomass. Note that the growth rates of the state variables will depend on the area of the swamp, which is subject to change due to processes of growth and decline, as well as on nutrients and other variables not identified explicitly. The dynamics of these variables, in turn, may be affected by management decisions as well as by the dynamics of the state variables, providing nonlinear linkages whose complexity increases with the number of variables considered.

22.3 The Necessity of Simplification

Even supposing that we can determine the general forms of such systems, we still would not be prepared to make management choices until we determined the growth rate and capacity functions $g(.)$ and $C(.)$. The standard statistical method for estimating such functions involves cross sections; that is, one observes many different swamps with different state variable configurations and uses regression methods to identify variables. Unfortunately, this procedure is rarely useful for ecological systems because they are so idiosyncratic – the relationships that characterize one swamp may not generalize at all to another, so the common structure necessary for cross-sectional analysis is missing. The difficulties of cross-sectional analysis of ecosystems is discussed in detail in Cole et al. (1991). Approaches to these problems include long-term observation and ecosystem experiments (see, for example, Pace and Groffman 1998). Alternatively, considerable progress has been made in mechanistic, individual-based approaches to assembling ecological systems (Levin et al. 1997; DeAngelis and Gross 1992), although dimensional reduction remains an essential and difficult challenge. Such individual-based models begin from known biophysical relationships and allow systems to develop, producing ensembles of possible outcomes. It is essential in such approaches therefore to determine what dynamical features are robust and to develop statistical methods to extract regularities from masses of computer runs. To that end, statistical mechanical methods sometimes can be developed to reduce the complexity of the simulation models to the essential core (Bolker et al. 2000; Levin and Pacala 1997 and Pacala and Levin 1997).

The mangroves are not unique in their complexity; indeed, they probably represent ideal starting points for investigation just because the issues are simpler than they are for more extended ecosystems. For example, in modeling the Pacific Continental Shelf, we must include whales, kelp, plankton, sea otters, shellfish, finfish, seals, and birds intertwined in their dynamics through complex and intricate webs of interaction. Detailed modeling of such systems that attempted to account for every component would be at best misguided and at worst seriously misleading. One must seek out methods of simplification, choosing the scales of space, time, and organizational complexity that allow us to focus on the most important variables involved in the associated choice. The need for simplification applies to any ecosystem including, of course, the mangroves.

Simplification can take many forms, and appropriate steps will depend on the system considered as well as the questions being asked. Depending upon circumstances, one may treat some variables (such as the total area of the swamp) as being constant and more generally ignore the dynamics of slow variables (such as mangrove regeneration). This might enable us to focus on the dynamics of the fish equations and on valuing losses in protection from storms and erosion as well as other public services.

In this way, we are often able to cull out the most important factors in a management problem. But progress will depend on obtaining good characterizations of the dynamics of the fast variables; this is no easy feat for ecological systems. Even for the dynamics of fisheries, which have been the object of intense study because of their economic importance, considerable uncertainties remain. Many of these may be traced to particular critical stages in the process, such as the recruitment of new individuals; but due to the nonlinearities inherent in these systems, small uncertainties in some parameters can become magnified through the loop of interactions, leading to large uncertainties in system dynamics. In such circumstances, the best strategy may be to be precautionary and adaptive (Holling 1978) – move in small steps, using continual monitoring to adjust harvesting and other management strategies. Thus, we could allow a small shrimp farm at the outset, monitor soil erosion and fish stocks, and proceed with a larger farm if the harms turn out to be small. Such adaptive policies are in general preferable to a "no change" strategy because they allow information gain through "adaptive probing" (Ludwig and Hilborn 1983). In contrast, policies of no change do not allow us to learn about changes in character the system may be experiencing far from the particular equilibrium; they further necessarily assume that the status quo is good, which may not be the case.

The approach described above is a highly aggregated one, ignoring the power of individual incentives. Changes in individual behaviors, in the aggregated model, do affect rewards but in diffuse ways that dissipate incentives. In dealing with problems of the commons, in general, we must find ways to tighten feedback loops, increasing individual rewards and internalizing externalities (Levin 1999, Heal 1996). Such approaches provide public incentives to users of the ecosystems. For the fishery, this might mean the use of tradeable quotas and common property resource management that would both serve to limit the catch and to provide incentives for using least-cost methods of fishing.

Typical permit systems in fisheries, called individual transferable quotas, allocate percent shares of an annually defined (variable) total allowable catch (TAC) rather than a fixed tonnage. This approach provides an adaptive management mechanism by allowing the TAC to be redefined annually in response to changing ecological conditions. The use of individual transferable quotas (ITQs) also reduces overcapitalization in the fishery, releasing the capital for more productive uses. It provides an accurate valuation of the fishery, one necessary component in valuing its interdependency with the mangroves. These systems are now used with considerable success around the world (National Research Council, Ocean Studies Board 1999).

Simplification through elimination of slow time scale dynamics can be helpful in many situations; but ultimately, there may be a price to pay if one does not consider the coupling between dynamics on multiple time scales. Ludwig et al. (1978) demonstrated the importance of such phenomena in their consideration of the dynamics of the spruce budworm in Canadian forests. The dynamics of slow variables hold the key to resilience and potential domain shifts. Although identification of such variables and potential loss of resiliency is problematical, their importance is fundamental to management.

22.4 Discontinuities and Multiple Time Scales: The Case of Temperate Lakes

Precautionary and adaptive approaches are well-justified techniques for dealing with uncertainty. Unfortunately, even the most cautious of adaptive management policies cannot protect against the potential for catastrophic changes in system character. It has long been a dream of theoreticians and managers alike to be able to detect hints of impending system collapse in measures of current system performance, but efforts to do so have been no more successful than efforts to predict when the stock market will experience meltdowns. There are many examples where seemingly small changes in one place have quite large impacts on the overall system, and these may be irreversible. Fisheries provide prototypical examples of such potential disasters – overfishing, for example, can reduce fish populations below sustainable levels, engendering collapses that may be difficult or impossible to reverse. More generally, ecosystems typically respond non-linearly to perturbation; their supply of services may hardly appear to change with incrementally increasing human (or natural) impacts up until a point, whereupon the response can be dramatic and very refractory to efforts to reverse.

These dangers are well illustrated by the ecology of phosphorus-limited shallow lakes in the border area between the United States and Canada, but apply more generally to temperate lakes worldwide. As with mangrove swamps, an accurate model would need to keep track of many variables: fish populations, phosphorus suspended in the water, algal blooms, and other plant life, phosphorus embedded in the mud of the lake bottom, and more. However, much as for the mangrove swamp, a sensible way to begin is through the development of simplified models incorporating macroscopic and highly aggregated variables.

As an example, Carpenter et al. (in press) model the dynamics of a lake in terms of the store of phosphorus (x) suspended in algae, according to a relation of the form

$$dx / dt = a\text{-}bx + f(x) \tag{22.4}$$

Here a denotes phosphorus inputs from the watershed (sometimes referred to as the "loading"); b is the purification rate – the rate of loss of phosphorus per unit stock (through sedimentation, outflow, and sequestration in organisms other than algae); and $f(x)$ is internal loading. This function is assumed to be "S-shaped"; that is, for low stocks of phosphorus, additions tend to be stored in the lake bed so that there is a relatively low marginal return to the water. For higher stocks this marginal return increases, only to fall again when maximal suspension is approached. This equation represents a "minimal state variable" approximation to the complex food web of a real lake. Carpenter et al. (in press) provide a justification for this abstraction, including a rationale for the form of the "recycling curve", $f(x)$. The recycling curve reflects natural positive feedback processes that are triggered inside a typical lake when phosphorus load becomes too high.

In the absence of external loading (a = 0), this system essentially has three behaviors (Fig. 22.1). At low purification rates (low b), the only stable equilibrium is a eutrophic one, corresponding to a high level of phosphorus. In this state, the water would be turbid even without external loading due to natural processes.

At the opposite extreme of high purification (high b), it is clear from Fig. 22.1 that natural processes will completely clear phosphorus from the system. As external loading (a) is increased, effectively shifting the still-sigmoidal total loading curve upward until it crosses the purification curve, a range of loading levels will be entered

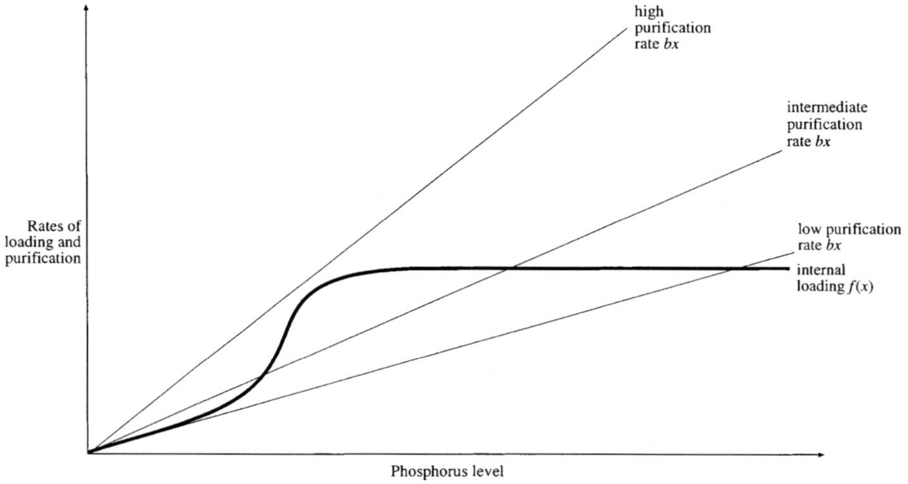

Fig. 22.1 Phosphorus loading and purification rates at three levels of purification. Intersection points indicate equilibria. Note that there are three possible equilibria at intermediate purification rates.

in which two stable states exist – an oligotrophic (clear water) state, which is highly valued by the lake's users, and a eutrophic one, which is disliked by the lake's users but indirectly valued by agricultural interests because it allows them high fertilizer use. Which state is reached for a given constant loading depends on the initial stock of phosphorus, with high initial stocks leading to a eutrophic long-run equilibrium. Note that all long-run outcomes are reversible in this case – sufficiently low loading rates will always restore the clear-water state if it is maintained for a sufficiently long time. However, the economic barriers to reversibility, for example, as agricultural interests become dependent on the high rate of fertilizer use, may be substantial.

If the external loading a is increased yet further (still in the case of high purification rate), a second threshold level will be reached where the oligotrophic equilibrium vanishes and only the eutrophic equilibrium remains. Even if the system was previously nestled comfortably in an oligotrophic state, the increase in loading can cause a jump to the eutrophic state. As long as the purification rate is high, however, this state (and hence that described in the previous paragraph) is reversible – reduction in the loading rate (a) will restore the system to a stable oligotrophic equilibrium whose clear water can be enjoyed by users.

This reversibility of eutrophy is lost, however, at intermediate purification rates (b). In this case, even without external loading, the system has two possible stable states – an oligotrophic one and a eutrophic one. In this case, like that described in the previous paragraph, increased loading can suddenly flip a system from oligotrophy to eutrophy. The difference is that now reduction of loading back to low levels will leave the system in a eutrophic state, since the high internal loading rates associated with eutrophication are now sufficient (relative to the intermediate purification rate) to maintain this condition. In this case, even very cautious adaptive management may fail to preserve the system in its oligotrophic state since indicators of impending qualitative shifts are in general lacking in such situations. This case emphasizes the point made by Arrow and Fisher (1974) that, in the face of irreversibilities, precaution in making additional development investments is advisable lest they turn out in retrospect to have been inappropriate. It may argue for the imposition of "safe minimum standards" [e.g., Berrens et al. (1998)] as a measure to maintain the system far from qualitative shifts.

22.5 Property Rights and Externalities

If one firm were to have the sole ownership of all the services provided by the mangroves or lakes, it could sell these services on markets and in effect operate as would any multiproduct firm (such as a farm). As a result, the ensuing management of the mangrove forest or the lake would be efficient; if the firm further were operating in competitive markets, it would also maximize the nominal social well-being. However, one important characteristic of these systems is that it is in general impossible to privatize all the services (i.e., assign private property rights to them) and that in practice most services are not assigned property rights.

Another example is provided by an open access fishery. Quite often, coastal fisheries supported by mangrove swamps or lake fisheries are open access; that is, anyone can use the fishery. This corresponds to a situation when there are no property rights whatsoever to the fishery. In this case, any fisherman will enter the fishery until all rents accruing to the fish stocks have been dissipated. Note that the net social revenue from the fishery is equal to the rent generated by it. Thus, with open access, the fishery does not provide any social good, and one component of the services provided by the mangrove forest has simply the value zero. In this case, the prime objective of the coastal or lake management must be to limit the access to the fishery in order to let the fish stocks recover and thereby generate a positive rent. ITQs, as mentioned earlier, provide one possible mechanism, but only one.

22.6 Discussion

Managing ecological systems raises challenges both old and new. In principle, the issues involved are those that apply to the management of any capital asset, but the relative importance of these is changed. Management involves in a fundamental way elements that may exist in traditional economic approaches but are given inadequate attention. In particular, traditional markets may not adequately reflect externalities and social costs. Furthermore, ecological systems are highly nonlinear, constrained by historical behavior, and with the potential for dramatic shifts in dynamics. The greatest challenge perhaps is in the valuation of the manifold services ecosystems provide humanity and in maintaining the resiliency that sustains them. To this end, we recommend precautionary and adaptive approaches coupled with mechanisms to tighten cost and benefit loops and to internalize externalities, including local empowerment and common property resource management.

The management of ecological systems is neither beyond the reach of economic analysis nor neatly treated within conventional frameworks. The particular challenges faced can lead to new dimensions of economic theory and new partnerships between economists and ecologists. The path is not straightforward, but the journey is one well worth making.

Acknowledgements The authors gratefully acknowledge the suggestions of Scott Farrow and three referees.

References

Arrow, K. J., & Fisher, A. C. (1974). Preservation uncertainty and irreversibility. *Quarterly Review of Economics, 87*, 312–319.

Berrens, R. D., et al. (1998). Implementing the Safe Minimum Standard approach: Two case studies from the U.S. Endangered Species Act. *Land Economics, 74*, 147–161.

Bolker, B. M., Pacala, S. W., & Levin, S. A. (2000). Moment methods for stochastic processes in continuous space and time. In U. Dieckmann, R. Law, & J. A. J. Metz (Eds.), *Geometry of ecological interactions: Simplifying spatial complexity*. Cambridge: Cambridge University Press.

Carpenter, S. R., Ludwig, D., & Brock, W. A. (in press) Management of eutrophication for lakes subject to potentially irreversible change. *Ecological Applications*.

Cole, J. J., Lovett, G., & Findlay, S. G. (1991). *Comparative analysis of ecosystems*. Berlin: Springer.

Costanza, R. R., et al. (1997). The value of the world's ecosystem services and natural capital. *Nature, 387*, 253–260.

Daily, G. C. (Ed.). (1997). *Nature's services: Societal dependence on natural ecosystems*. Washington, DC: Island Press.

DeAngelis, D. S., & Gross, L. J. (1992). *Individual-based models and approaches in ecology: Populations, communities and ecosystems*. New York: Chapman & Hall.

Heal, G. (1996). Markets and sustainability. In Stewart, R. (Ed.) *Conference Proceedings on Environmental Governance at La Pietra, Florence, July 1996*. Cambridge University Press: (Submitted)

Holling, C. S. (Ed.). (1978). *Adaptive environmental assessment and management* (p. 377). New York: Wiley.

Levin, S. A. (1998). Ecosystems and the biosphere as complex adaptive systems. *Ecosystems, 1*, 431–436.

Levin, S. A. (1999). *Fragile dominion: Complexity and the commons*. Reading, MA: Perseus Books.

Levin, S. A., Grenfell, B., Hastings, A., & Perelson, A. S. (1997). Mathematical and computational challenges in population biology and ecosystems science. *Science, 275*, 334–343.

Levin, S. A., & Pacala, S. W. (1997). Theories of simplification and scaling of spatially distributed processes. In D. Tilman & P. Kareiva (Eds.), *Spatial ecology: The role of space in population dynamics and interspecific interactions* (pp. 271–296). Princeton, NJ: Princeton University Press.

Ludwig, D., & Hilborn, R. (1983). Adaptive probing strategies for age structured fish stocks. *Canadian Journal of Fisheries and Aquatic Sciences, 40*, 559–569.

Ludwig, D., Jones, D. D., & Holling, C. S. J. (1978). Qualitative-analysis of insect outbreak systems: Spruce budworm and forest. *Journal of Animal Ecology, 47*, 315–332.

National Research Council, Ocean Studies Board. (1999). *Sharing the fish: Toward a national policy on individual fishing quotas*. Washington, DC: National Academy Press.

Pacala, S. W., & Levin, S. A. (1997). Biologically generated spatial pattern and the coexistence of competing species. In D. Tilman & P. Kareiva (Eds.), *Spatial ecology: The role of space in population dynamics and interspecific interactions* (pp. 204–232). Princeton, NJ: Princeton University Press.

Pace, M. L., & Groffman, P. M. (1998). *Successes, limitations and frontiers of ecosystem science*. Berlin: Springer.

Chapter 23
Askö in Washington 1999: Commentary by Brian Walker

Brian Walker

"Managing" resources in the context of this paper amounts to dealing with the trade-offs in choices that managers have to make. In assessing these choices, it is necessary to use some valuation of the costs and benefits involved, and the key message from this paper is that any such valuation must take into account several important ecological aspects of ecosystems that in the past have been too often overlooked. To quote the authors: "The exploitation of natural systems, including efforts to modify and manage them, forces us to confront the tradeoffs between real and potential services and the effects upon their resiliency" (p. 1401).

A major difference in the ways ecologists and economists have approached ecosystem management is that ecologists have tended to confront and highlight complex dynamics in the ecosystem, while economists have sought principles that lead to increases in net social welfare resulting from the use of the ecosystem. The authors point out the features of ecosystems that make good "efficient" (in the economic sense) management inherently problematic: They are highly non-linear, exhibiting multiple domains of attraction; their development to their current states is highly idiosyncratic and historically constrained; ecological assets in large part represent public goods with consequent externalities that are not dealt with in market mechanisms.

There are several important messages that ecologists do not usually consider, and I highlight them here in order to aid a better understanding by ecologists of what economists are trying to achieve. First, the need to understand the welfare function for social well-being that economists use (see Eq. 22.1). It equates to the utility society derives from the full array of resource uses (consumption), and this aggregated utility function is typically concave, meaning that it reflects diminishing returns. This means that its assessment requires assigning some discount rate, and determining the appropriate discount rate is highly problematical, involving choices about intergenerational equity. The authors (no doubt influenced by the ecologists

B. Walker
CSIRO Sustainable Ecosystems, Canberra, Australia
e-mail: brian.walker@csiro.au

T. Söderqvist et al. (eds.), *Bringing Ecologists and Economists Together*,
DOI 10.1007/978-90-481-9476-6_23,

among them) make a very important point that is easy to pass by. They point out that optimal management amounts to a choice of all the control variables that determine the future development of the system – a choice that maximizes social welfare; *and* that the choice need not be fixed. In adaptive management approaches (as advocated by ecologists) continual updating "may be necessary" (authors' words, p. 1402). Given the inherent uncertainties in ecosystem dynamics, and in their responses to management interventions, an adaptive management approach is a necessary component of successful ecosystem management; and therefore any utility function and its associated choice of controls needs to be regularly reviewed and updated. Walters (1997) considers that discounting is an important reason why attempts to apply adaptive management have often failed, and Ludwig et al. (2005) show that proper analysis of uncertainty for the discount rate leads to lighter discounting of the future, which is favorable for adaptive management. So if we apply the same stringency to economic uncertainties that we apply to ecological uncertainties, the formalism in this paper by Arrow et al. leads to policies that favor adaptive ecosystem management, indicating that economics and ecology are in agreement about the importance of adaptive management. So for these reasons I would emphasize that the choice of interest rate (and other controls of the welfare function) need to be regularly updated, rather than just "may be necessary".

The paper presents a simplified model of a mangrove system in which fish and mangrove trees are harvested. It is used to illustrate the need for simplification if strategies are to be assessed, and offers some insights into how this might be achieved (like initially holding slow variables constant), and tying this into a precautionary, adaptive strategy of small intervention steps coupled closely to monitoring. The slow variable discussion leads to the importance of likely threshold effects on such variables, and the significance of alternate stability domains. Using the example of shifts in lake regimes due to phosphorous loading, the conclusion is again reached that a precautionary, adaptive approach is the best way to proceed. There is an interesting comment, viz. "It has long been a dream of theoreticians and managers alike to be able to detect hints of impending system collapse in measures of current system performance, but efforts to do so have been no more successful than efforts to predict when the stock market will experience meltdowns" (p. 1404). Using the example of regime shifts in lakes, the authors conclude that "…even very cautious adaptive management may fail…since indicators of impending qualitative shifts are in general lacking in such situations" (p. 1405). While this may be true, in general, it is encouraging to note that since the paper was written there has been some advance in ecosystem understanding. For example, a paper by Carpenter and Brock (2006) on rising variance as a leading indicator of a regime shift in such lakes offers some hope that, in ecosystems at least, there is a chance that we may learn how to detect thresholds before crossing them.

The perspective in this paper is one of managers outside the system they are managing. The dynamics of interest are all in the ecosystem. How these dynamics might play out differently over time if the dynamics of the social system (of the managers and others) were also included would be interesting to know. Groups of ecologists, economists and social scientists are attempting to understand the combined dynamics of linked social-ecological systems (see, for examples, www.resalliance.org) and the questions of interest, in the context of this paper, would have to do with the consequences of a "flip" in the mangrove or the lake for the dynamics of the social system. Could it, too, undergo a "flip", as a result of the changes in the ecosystem, and how might this in turn feed back to changes in ecosystem management? Understanding and managing regime shifts in social-ecological systems is all about understanding changes in feedbacks associated with threshold effects. We seem to be further advanced in incorporating non-linear dynamics of ecosystems in our assessment than we do the dynamics of social systems. Incorporating both in simple models is a difficult challenge.

The paper concludes with some sound advice. The inability to assign property rights to many ecosystem services, and the fact that even where possible most services are not assigned property rights, means that management of most ecosystems is, in an economic sense, inherently inefficient. Social wellbeing cannot therefore be maximized by relying on only market mechanisms. In the view of the ecologists and economists who wrote this paper these difficulties call for new dimensions of economic theory and "new partnerships between economists and ecologists" (p. 1405). And while this is developing, the inadequacy of traditional markets and the non-linear behavior of ecosystems call for precautionary and adaptive approaches. Ecological readers should be encouraged by the willingness of eminent economists to acknowledge these needs, and to engage in the collaboration.

References

Carpenter, S. R., & Brock, W. A. (2006). Rising variance: A leading indicator of ecological transition. *Ecological Letters, 9*, 311–318.

Ludwig, D., Brock, W. A., & Carpenter, S. R. (2005). Uncertainty in discount models and environmental accounting. *Ecology and Society, 10*(2), 13. [online] URL: http://www.ecologyandsociety.org/vol10/iss2/art13/.

Walters, C. J. (1997). Challenges in adaptive management of riparian and coastal ecosystems. *Conservation Ecology, 1*(2), 1. [online] URL: http://www.consecol.org/vol1/iss2/art1/.

Chapter 24
Askö in Washington 1999: Commentary by Edward B. Barbier

Edward B. Barbier

Since the paper by Arrow et al., various studies have also focused on the critical need to "value" ecosystem services and the different methods for conducting such valuation (e.g., see Freeman 2003; NRC 2005; Pagiola et al. 2004). As a report from The US National Academy of Science has emphasized, "the fundamental challenge of valuing ecosystem services lies in providing an explicit description and adequate assessment of the links between the structure and functions of natural systems, the benefits (i.e., goods and services) derived by humanity, and their subsequent values" (NRC 2005, p. 2). Consideration of uncertainty and irreversible loss are perhaps important issues, but as emphasized by this report as well as in Arrow et al., perhaps a more "fundamental challenge" in valuing these flows is that ecosystem services are largely not marketed. Unless some attempt is made to value the aggregate willingness to pay for these services, then management of natural ecosystems and their services will not be efficient.

To illustrate the latter "challenge", Arrow et al. draw extensively on the example of mangrove ecosystems, but without providing actual valuation estimates of some of the more important services of these critical coastal ecosystems, such as habitat-fishery linkages and storm protection. However, recent valuation studies do provide evidence in support of the authors' views on these important but often neglected ecosystem services. The rest of this commentary will discuss some empirical results, using the example of mangroves in Thailand.

Since 1961 Thailand has lost from 1,500 to 2,000 km^2 of coastal mangroves, or about 50–60% of the original area (FAO 2003). Over 1975–1996, 50–65% of Thailand's mangroves were lost to shrimp farm conversion alone (Aksornkoae and Tokrisna 2004). The primary cause of the mangrove loss has been shrimp aquaculture expansion, as well as other coastal developments (Barbier and Sathirathai 2004).

An important issue is whether the loss of these mangroves has resulted in substantial welfare losses for Thailand as a consequence. If the foregone benefits of the

E.B. Barbier
Department of Economics and Finance, University of Wyoming, USA
e-mail: ebarbier@uwyo.edu

T. Söderqvist et al. (eds.), *Bringing Ecologists and Economists Together*,
DOI 10.1007/978-90-481-9476-6_24,

ecological services of mangroves are not large, then perhaps mangrove deforestation has not been a serious problem. Based on previous studies of mangrove benefits in Thailand, Table 24.1 indicates the value of three of these benefits: the net income from local mangrove forest products, habitat-fishery linkages and storm protection. The table suggests that the loss of these benefits from mangrove deforestation is likely to be large for Thailand.

The easiest ecosystem value to estimate, which is the value to local communities of using mangrove resources in terms of the net income generated from the forests in terms of various wood and non-wood products, is actually the lowest of the three at around $101 per ha. Based on a type of intertemporal bioeconomic fishing model similar to that proposed by Arrow et al., the habitat-fishery linkage yields a net annual benefit of $249 per ha. Finally, by using an "expected damage function" approach, Barbier (2007) estimates the storm protection service of mangroves to be $1,879 per ha. The total annual sum of these three mangrove benefits is $2,229 per ha in constant 1996 prices. Although the value of the storm protection service clearly dominates these benefits, all three are significantly higher than zero. In addition, the net present value of all three mangrove ecosystem benefits ranges from $16,046 to $20,876 per ha.

In sum, Arrow et al. are correct to emphasize that the "efficient management of ecosystems involves the same economic principles as does efficient management of fossil fuels and other capital assets" (p. 1401). The problem is that, because the important services of ecosystems are generally non-marketed, the value of these assets is often "underpriced" in development decisions that lead to their irreversible conversion or damage. In recent years substantial progress has been made by economists working with ecologists and other natural scientists on the "fundamental challenge" to improve environmental valuation methodologies applied to ecosystem services. Now it is up to policymakers to take such valuations seriously in their development decisions.

Table 24.1 Net present value of mangrove forest benefits, Thailand (1996 US$)[a]

	Value (US$)/ha
Net income from timber and non-timber products[b]	101.49
Habitat-fishery linkages[c]	248.70
Storm protection[d]	1,878.98
Total benefits	2,229.17
Net present value (10% discount rate)	20,876.00
Net present value (12% discount rate)	18,648.74
Net present value (15% discount rate)	16,046.08

[a]All benefits estimated on an annual basis; net present value calculations are based on a 20-year time horizon
[b]Adapted from Sathirathai and Barbier (2001)
[c]From Barbier (2003), assuming a price elasticity of demand for fish of –0.5
[d]From Barbier (2007)

References

Aksornkoae, S., & Tokrisna, R. (2004). Overview of shrimp farming and mangrove loss in Thailand. In E. B. Barbier & S. Sathirathai (Eds.), *Shrimp farming and mangrove loss in Thailand*. London: Edward Elgar.

Barbier, E. B. (2003). Habitat-fishery linkages and mangrove loss in Thailand. *Contemporary Economic Policy, 21*, 59–77.

Barbier, E. B. (2007). Valuing ecosystem services as productive inputs. *Economic Policy 22*, 177–229.

Barbier, E. B., & Sathirathai, S. (Eds.). (2004). *Shrimp farming and mangrove loss in Thailand*. London: Edward Elgar.

FAO. (2003). Status and trends in mangrove area extent worldwide. In M. L. Wilkie, & S. Fortuna (Eds.), Forest Resources Assessment Working Paper No. 63. Rome: Forest Resources Division, Food and Agricultural Organization of the United Nations.

Freeman, A. M., III. (2003). *The measurement of environmental and resource values: Theory and methods* (2nd ed.). Washington, DC: Resources for the Future.

National Research Council (NRC) (2005). *Valuing ecosystem services: Toward better environmental decision making*. Washington, DC: The National Academies Press.

Pagiola, S., von Ritter, K., & Bishop, J. (2004). *How much is an ecosystem worth? Assessing the economic value of conservation*. Washington, DC: The World Bank.

Sathirathai, S., & Barbier, E. (2001). Valuing mangrove conservation, Southern Thailand. *Contemporary Economic Policy, 19*, 109–122.

Chapter 25
Askö in Stanford 2000: Are We Consuming Too Much?*

Kenneth Arrow, Partha Dasgupta, Lawrence Goulder, Gretchen Daily, Paul Ehrlich, Geoffrey Heal, Simon Levin, Karl-Göran Mäler, Stephen Schneider, David A. Starrett, and Brian Walker

Is humanity's use of Earth's resources endangering the economic possibilities open to our descendants? There is wide disagreement on the question. Many people worry about the growth in our use of natural resources over the past century. Some of this increase reflects the higher resource demands from a growing world population. But it also reflects the growth of per capita output and consumption. During the twentieth century, world population grew by a factor of 4 to more than 6 billion, and industrial output increased by a factor of 40. Per capita consumption in industrialized nations today is far higher than it was 100 years ago, and some would argue that this is irresponsible in the light of its implications for resource demands.

---

*This article was published in *Journal of Economic Perspectives*, Vol. 18, Arrow et al., Are we consuming too much?, 147–172, Copyright American Economic Association (2004).

K. Arrow, L. Goulder, and D.A. Starrett
Department of Economics, Stanford University, Stanford, CA 94305, USA

P. Dasgupta
Department of Economics, University of Cambridge, Cambridge CB3 9DD, UK

G. Daily, P. Ehrlich, and S. Schneider
Department of Biological Sciences, Center for Conservation Biology, Stanford University, Stanford, CA 994305, USA

G. Heal
Columbia Business School, Uris Hall, 3022 Broadway, New York, NY 10027-6902, USA

S. Levin
Department of Ecology and Evolutionary Biology, Princeton University, Princeton, NJ 08544, USA

K.-G. Mäler
Beijer Institute of Ecological Economics, the Royal Swedish Academy of Sciences, Box 50005, SE-104 05 Stockholm, Sweden

B. Walker
CSIRO Sustainable Ecosystems Division, Canberra, Australia

T. Söderqvist et al. (eds.), *Bringing Ecologists and Economists Together*,
DOI 10.1007/978-90-481-9476-6_25, © Springer Science+Business Media B.V. 2011

In the last 100 years, energy use has increased by a factor of 16, annual fish harvesting by a multiple of 35 and carbon and sulfur dioxide emissions by a factor of 10. The application of nitrogen to the terrestrial environment from human use of fertilizers, fossil fuels and leguminous crops is now at least as great as that from all natural sources combined (McNeill 2000). If we look at specific resources and services, such as fresh water, the atmosphere as a carbon sink, and a wide variety of ecosystem services, evidence suggests that continuing growth in their utilization rates is unsustainable (Vitousek et al. 1986; 1997; Postel et al. 1996).

On the other hand, it may be claimed that, just as earlier generations invested in capital goods, research and education to bequeath to current generations the ability to achieve high levels of consumption, current generations are making the investments that are necessary to assure higher real living standards in the future, despite stresses on the natural resource base. Indeed, historical trends in the prices of marketed natural resources and the recorded growth in conventional indices of economic progress in currently rich countries suggest resource scarcities have not bitten as yet (Barnett and Morse 1963; Johnson 2000). This optimistic viewpoint emphasizes the potential of capital accumulation in the form of increased manufactured capital and human capital, as well as technological change, to compensate for the diminishment of natural resources.

This paper, the outgrowth of discussions among a group of ecologists and economists, offers an analysis that we hope will go some way toward reconciling the conflicting intuitions. The binocular vision that can be obtained from using both ecological and economic insights raises questions that might not occur in either viewpoint alone.

By what criterion should we judge whether consumption is or is not excessive? Economic analysis has tended to emphasize the criterion of maximizing the present discounted value of current and future utility from consumption – what we will call *intertemporal social welfare*. Current consumption is deemed excessive or deficient depending on its level relative to that called for by this optimization problem. However, analysts with a different perspective have applied a criterion of *sustainability*, which emphasizes the ability of the economy to maintain living standards. In this paper, we will examine these criteria for evaluating whether consumption is excessive and identify factors in the economic and ecological domains that determine whether or not it is in fact so.

25.1 Alternative Criteria for the Evaluation of Consumption

In the framework presented here, the underlying elements of intertemporal social welfare are consumption (broadly defined) and utility. Then the intertemporal social welfare V_t at time t can be defined as the present discounted value of the flow of utility from consumption from the present to infinity, discounted using the constant rate δ (>0).[1] In focusing at each moment on aggregate consumption, our framework abstracts from *intra*temporal equity issues. The equity issues that preoccupy us are intertemporal.

One of the determinants of V_t is the "productive base," which consists of society's capital assets and institutions at t. The capital assets include manufactured capital, human capital and natural capital. The productive base also includes the knowledge base and society's institutions. Although institutions are frequently regarded to be capital assets themselves, we will instead view institutions as guiding the allocation of resources – including capital assets. Institutions include the legal structure, formal and informal markets, various agencies of government, interpersonal networks and the rules and norms that guide their behavior. In what follows, we will generally employ the term "technology" to refer to both the knowledge base and the institutions that influence, among other things, the effectiveness with which this knowledge is put to use. We now discuss two criteria for judging whether or not current consumption is excessive.

25.1.1 The Maximize Present Value Criterion

One can think of intertemporal social welfare V_t as a function of initial conditions – the capital assets and level of technology at time t – and the choices from time t forward as to how to allocate output between investment and consumption. According to the *maximize present value criterion*, actual consumption today is excessive if it is greater than the level of current consumption prescribed by this optimal consumption path. To put it another way, current consumption is excessive if lowering it and increasing investment (or reducing disinvestment) in capital assets could raise future utility enough to more than compensate (even after discounting) for the loss in current utility.

The optimal path depends, among other things, on the discount rate, δ. A higher value for δ, other things being equal, means that less weight is placed on future

[1] Let s and t variously denote time (where $s \geq t$). Let $C(s)$ denote a society's aggregate consumption and $U(C(s))$ the flow of utility at time s. Marginal utility is assumed to be positive. Let V_t denote intertemporal social welfare at time t, defined as the present discounted value of the flow of $U(C(s))$ from t to infinity, discounted using the constant rate δ (> 0). Assuming continuous time, we have

$$V(t) = \int_{s=t}^{\infty} U[C(s)]e^{-\delta(s-t)}ds.$$

In calling U "utility," we are not necessarily subscribing to classical utilitarianism. We are assuming, more generally, that consumption has a social worth, which we call utility $U(C)$, and that V_t is a numerical representation of an ethical ordering over infinite utility streams beginning at t. Koopmans (1972) has identified conditions on orderings that permit one to express V_t in the form we are using here.

utility. The "right" value of δ has long been a matter of debate. Ramsey (1928) argued that in a deterministic world the appropriate value of δ is zero, implying that the utility of future people ought to receive the same weight as that of people today. Koopmans (1960), however, showed that applying a zero social rate of pure time preference can lead to paradoxes. Lind (1982) and Portney and Weyant (1999) contain diverse arguments concerning the appropriate choice of δ.

One interesting aspect of the optimal consumption path according to the maximize present value criterion is that it can be linked, theoretically, to the outcome of a decentralized market economy. In a fully competitive economy with a complete set of futures markets and no externalities, and in which individuals discount their future utility at the social rate δ, the time path of consumption will correspond to the optimal consumption path. Conversely, in a world where markets for many future goods and many types of risk bearing do not exist and where environmental externalities prevail, consumption generally will not be optimal. Nevertheless, it is possible to conduct social cost–benefit analysis to judge whether a policy reform at t increases intertemporal social welfare (Dasgupta 2001a; Arrow et al. 2003b).

25.1.2 The Sustainability Criterion

An alternative yardstick for evaluating time profiles of consumption is the *sustainability criterion*. The terms "sustainability" and "sustainable development" became commonplace after the report by the WCED (1987), widely known as the Brundtland Commission (after its chairperson). Sustainable development was defined as "development that meets the needs of the present without compromising the ability of future generations to meet their own needs." Several interpretations of sustainability are compatible with this phrase (Pezzey 1992; Solow 1992; Heal 1998; Asheim 2003). Here we take sustainability to mean that intertemporal social welfare V must not decrease over time. Thus, we will say that the sustainability criterion is satisfied at time t if $dV_t/dt \geq 0$.

Several features of this criterion for sustainability deserve emphasis. First, the criterion concentrates on the change in V, not on V's level. Second, even if a consumption path were to satisfy the criterion today and at all future dates, it would not guarantee that utility (U) in each future moment will be as high as it is today (Asheim 1994). Third, the criterion does not identify a unique consumption path: the criterion could in principle be met by many consumption paths. Fourth, if exhaustible resources are sufficiently important in production and consumption, then it is conceivable that no sustainable development program exists (Dasgupta and Heal 1979, chapter 7). Fifth, satisfying the sustainability criterion today does not guarantee that the criterion will be satisfied at all future times: a given consumption path may imply a rising V over the interval from now to the next period, but a falling V when it is evaluated over some future interval of time. Sixth, in defining sustainable development, there is no presumption that the consumption path being followed is optimal in the sense of maximizing V. And seventh, a consumption path that satisfies the sustainability criterion need not be intertemporally efficient; an alternative, Pareto-superior path may well exist.

The notion of society's "productive base" is closely connected to the issue of whether the sustainability criterion is satisfied. As defined earlier, the productive base is the stock of all society's capital assets at *t*, inclusive of manufactured capital assets, human capital and natural capital. It also depends on the level of technology. For now, we will regard the level of technology as fixed; we will consider technological change below.

Let *genuine investment* refer to the change in the productive base. Genuine investment can be expressed as the sum of the values of investments or disinvestments in each of society's capital assets, where the value of each investment is the product of the change in the quantity of the asset times the shadow value or accounting price of that asset.[2] Clearly, intertemporal welfare $V(t)$ is nondecreasing at t if and only if genuine investment is non-negative at t.[3] We can refer to genuine investment as the change in society's *genuine wealth*.

This requirement that the productive base be maintained does not necessarily entail maintaining any *particular* set of resources at any given time. Even if some resources such as stocks of minerals are drawn down along a consumption path, the sustainability criterion could nevertheless be satisfied if other capital assets were accumulated sufficiently to offset the resource decline.

While the criterion for sustainable development is straightforward to express, implementing it poses severe empirical challenges. Measuring changes in quantities of capital stocks is very difficult. It is especially hard in the case of stocks of natural resources such as minerals, fossil fuels, fish or insects. In evaluating the social losses from reductions in natural resources – and thus the alternative investments necessary to offset such losses – one must consider in principle all of the contributions of natural resources to present and future utility. Such contributions

---

[2] Let $\boldsymbol{K}_t$ denote the vector of stocks of all capital assets at t. Plainly, V is a function of $\boldsymbol{K}_t$. In the particular case where V is stationary (that is, where t itself does not directly influence V), we can write $V_t = V(\boldsymbol{K}_t)$. Let K_{it} denote the stock of the ith capital good at date t. By the chain rule of differentiation,

$$dV / dt = \sum_i (\partial V / \partial K_{it})(dK_{it} / dt) = \sum_i p_{it} I_{it},$$

where p_{it} ($\equiv \partial V/\partial K_{it}$) is the shadow price or accounting price of K_{it}, and I_{it} ($\equiv dK_{it}/dt$) denotes the rate of change in K_{it}. The right-hand side of the expression above is genuine investment.

It is frequently convenient to divide the expression for dV/dt by the value of some one kind of capital (valued at its shadow price). Thus, if there are just two kinds of capital, denoted, say, by K_1 and K_2, respectively, genuine investment measured relative to the first kind of capital is

$$(dV / dt) / (p_{K1} K_1) = (1 / K_1) dK_1 / dt + \left[p_{K2} K_2 / (p_{K1} K_1)\right](1 / K_2)(dK_2 / dt).$$

This indicates that our sustainability criterion can be expressed as the growth rate of one kind of capital plus the product of the growth rate of the other kind of capital and an adjustment factor. Note that, from the definition of shadow prices above, the adjustment factor is the elasticity of K_1 with respect to K_2 along an isoquant of $V(K_1, K_2)$. We apply this expression in a later section, where we extend the expression for dV/dt to account for technological change.

[3] This result was proved for optimally managed economies by Pearce and Atkinson (1995) and for arbitrary economies by Dasgupta and Mäler (2000).

may be direct as, say, objects of natural beauty; or they may be indirect, as in the contributions of ecosystem services such as water purification, flood control, climate stabilization, pollination of crops, control of agricultural pests and the generation and maintenance of soil fertility (Daily 1997); or they may be both (a wetland). Measuring these services is no easy task.

Another great challenge is determining how much more of one type of capital asset would be needed to compensate for the loss of one unit of another type of capital asset. Ecologists express concerns that natural resources have only a limited set of substitutes. They fear that to the extent that economists are overly optimistic about opportunities to substitute other capital assets for certain natural resources, reductions in their stocks would then receive too little weight. A major goal of ecological economics is to increase our understanding of the ways in which different kinds of natural capital contribute to human well-being and the extent to which they are substitutable for one another and for other kinds of capital assets.[4]

25.1.3 Extending the Sustainability Criterion to Account for Changing Population

In initial discussion of intertemporal social welfare, V, we have abstracted from the impact of changes in population size. Dealing with a changing population presents empirical challenges: forecasting the time profile of population is quite difficult. But the theoretical issues associated with a changing population may be even more challenging, as many of the theoretical problems remain unresolved. Should social welfare seek to maximize the sum of the utilities of all individuals, or should it seek to maximize some form of average welfare over all people – now and in the future? How can one compare the social value of a large population with higher total utility versus a smaller population with higher average utility?

One way to proceed is to regard population as another asset, in addition to the forms of capital we have already considered. This approach implies that for any given conception of intertemporal social welfare, there is an accounting price of population. Under this approach, genuine wealth is the accounting value of all capital assets, including population. The sustainability criterion continues to require that genuine wealth at constant accounting prices must not decline, but now invokes a broader notion of genuine wealth.

---

[4] On substitutability, see, for example, Dasgupta and Heal (1979, chapter 7), Ehrlich and Ehrlich (1990), Daily (1997), Daily et al. (2001), Levin (2001), Heal et al. (2001), and Heal (2003). An important complication related to assessing substitution possibilities is that the potential for substitution can vary by location. A natural resource in one place, like a local woodland, is not the same economic commodity as the same natural resource in another place. Dasgupta (1993) discusses the implications of local nonsubstitutability for the world's poorest people, who often have no substitutes available to them when their local resource base is degraded. For an interchange on the relative importance of substitution possibilities at aggregate and local levels, see Johnson (2001) and Dasgupta (2001b).

Of particular interest is the case where the intertemporal social welfare function is taken to be the discounted present value of the flow of total utility divided by the discounted present value of population size over time, where the discount rate used in both the numerator and denominator is the social rate of pure time preference, δ. This is a form of "dynamic average utilitarianism."[5] Dasgupta (2001a) has shown that if dynamic average utilitarianism represents intertemporal social welfare, then under certain conditions a nondeclining V_t is equivalent to the intuitively appealing requirement that *genuine wealth per capita* must not decline. The conditions in question are (i) population changes at a constant rate; (ii) per capita consumption is independent of population size (but presumably dependent on per capita capital assets); and (iii) all transformation possibilities among goods and services exhibit constant returns to scale. We shall make use of this finding when deriving certain empirical results later in this paper.

25.1.4 Extending the Sustainability Criterion to Account for Technological Change

In the presence of technological change, output and consumption could rise, and thus intertemporal social welfare V could increase, even if aggregate investment in terms of manufactured, human and natural capital were negative. How can we take such changes into account in assessing changes in intertemporal social welfare, V?

Here we enter uncharted territory, since there appears to be no prior literature that makes such a connection. We explore this relationship by examining how total factor productivity alters the assessment of V. In a rather special model – one with strong simplifying assumptions – we derive a formula for incorporating total factor productivity in V. Let γ represent the rate of growth of total factor productivity. γ is often called the "residual," since it is what remains after subtracting the influence of other factors on output growth. Let K (without subscript) represent a composite of society's natural, physical and human capital assets.[6] Let α be the elasticity of output with respect to K. Later in the discussion we will observe that genuine savings rates are close to 0. A straightforward argument shows that with zero rates of

[5] The formula for V in this case is

$$V(t) = \int_t^\infty N(s)U(c(s))e^{-\delta(s-t)}ds \Big/ \int_t^\infty N(s)e^{-\delta(s-t)}ds,$$

where $c(s)$ and $N(s)$ represent per capita consumption and population size, respectively, at time s. Notice that, to the extent that the denominator of the above form of V_t is not affected by policy, the maximize present value criterion implies the same optimum no matter whether "total" or "dynamic average" utilitarianism is adopted as the conception of intertemporal social welfare. However, it matters hugely whether "total" or "dynamic average" utilitarianism is adopted when one is applying the sustainability criterion. For further analysis of sustainability under changing population size, see Arrow et al. (2003a).

[6] K can also be regarded as a measure of genuine wealth prior to the adjustment in such wealth to account for technological change.

savings, total factor productivity growth at rate γ raises the growth rate of intertemporal social welfare (measured in units of capital) by γ/α.[7]

Thus, after calculating genuine investment and the associated change in genuine wealth, one can account for technological change by adding γ/α to the initially estimated growth rate of genuine wealth. It should be noted that there may be serious problems with published estimates of total factor productivity and its growth rate, γ, because national accounts do not include the economy's use of nonmarketed natural and environmental resources. Suppose that over a period of time, an economy extracts from its natural resource base at an increasing rate. Then the estimate of total factor productivity growth would be biased upward: some of the estimated growth in total factor productivity would actually be due to increased resource use rather than an increase in the knowledge base or improvements in institutional performance. On the other hand, if the rate of natural resource extraction were falling, recorded growth of total factor productivity would understate the true contribution of changes in the knowledge base. Our empirical investigation later in this paper relies on estimates of total factor productivity from models that do not consider natural resource extraction. Ideally, estimates of the productivity residual should derive from models that incorporate such extraction.[8]

25.1.5 Does Satisfying One Criterion Imply That the Other Is Satisfied?

The two criteria we have offered for assessing whether consumption is too high reflect different ethical considerations. An economic program that satisfies the sustainability criterion need not satisfy the maximize present value criterion. Conversely, an economic program that is optimal under the maximize present value criterion might not satisfy the sustainability criterion.

To see this, suppose that aggregate output is a Cobb-Douglas function of manufactured capital and the flow of an exhaustible natural resource. (Returns to scale may be either constant or decreasing.) Solow (1974) showed that sustainable development is technically feasible if the output elasticity with respect to manufactured capital exceeds the output elasticity with respect to the flow of the natural resource

[7] Assume that $K(t)$, the productive base, is a constant ($= K_0$) and that consumption = output = $A_0K_0^{\alpha}e^{\gamma t}$ for all t. From the definition of V in footnote 2, V is constant when $A_0K_0^{\alpha}$ is constant. Thus, the elasticity of A_0 (total factor productivity) with respect to K_0 along an isoquant of V is α^{-1}. In the second equation in footnote 2, substitute the productive base K for what was termed K_1 and total factor productivity A_0 for what was termed K_2. Since $(1/A)\ (dA/dt) = \gamma$, it follows from this equation that the sustainability criterion (expressed in terms of K) is the rate of growth of K plus γ/α.

[8] In an appendix to this paper available at <http://www.stanford.edu/~goulder>, we derive a formula for correcting the bias in estimating the rate of growth of total productivity induced by the omission of resources.

in production. Dasgupta and Heal (1979) showed, however, that so long as $\delta > 0$, consumption paths that are optimal in the sense of maximizing intertemporal welfare V involve consumption approaching zero in the long run; thus, they do not yield development that is sustainable over the long term.

The differences between the two criteria have potential implications for public policy. An investment project that passes the social cost–benefit test (namely, that its acceptance would increase today's V), could result in a decrease in V at some future date. Therefore, the standard policy remedies for improving economic efficiency – like establishing property rights, addressing externalities and so forth – do not guarantee sustainability.

25.2 Empirical Evidence Relevant to the Maximize Present Value Criterion

According to the maximize present value criterion, today's consumption is excessive if it is higher than the level of current consumption along the consumption path that maximizes the current intertemporal welfare V. No one can seriously claim to pinpoint the optimal level of current consumption for an actual economy. However, theoretical considerations can identify factors that would cause current consumption to be different in a predictable direction from the optimal level. Relevant theoretical issues include the relationship between market rates of return on investment and optimal social interest rates on consumption, and the relationship between market prices of contemporaneous goods (including those of current capital goods) and the social costs of those commodities.

25.2.1 The Market Rate of Return on Investment and the Social Rate of Interest on Consumption

For a consumption path in a market economy to be socially optimal, the market rate of return on investment, i, must be equal to the social rate of interest on *consumption*, denoted by r. If i exceeds r, markets are biased toward insufficient saving and excessive current consumption.

It can be shown that, if intertemporal social welfare is given by the form of V we have postulated here, the social rate of interest on consumption r is given by the relation $r = \delta + \eta g$, where δ is the social rate of pure time preference, η is the elasticity of marginal (social) utility, and g is the rate of growth in aggregate consumption.[9] The parameter δ reflects impatience, as is well known. The second

[9] For one derivation, see Arrow and Kurz (1970). For additional discussion, see Arrow et al. (1996).

term in the equation, η times g, may require explanation. One can view η as accounting for the fact that, to the extent that future generations have higher incomes, their consumption will be higher and the marginal utility of their consumption will be lower. However, η need not simply reflect diminishing marginal utility of consumption to an individual: it can be interpreted as a *social* preference for equality of consumption among generations. Thus, its function can be similar to the social rate of pure time preference, δ.

Does i exceed r? One might argue that this cannot be answered from empirical observation alone, on the grounds that the choices of δ and η are inherently value judgments – subjective matters. However, there exists a different approach that offers a link to empirical behavior. This approach asserts that a "typical" individual's preferences should guide our choices of δ and η. In particular, it identifies the social rate of pure time preference, δ, with the utility discount rate that a typical individual would endorse. The assumptions of this approach are clearly subject to debate, but it seems worthwhile to consider what they imply for the relationship between i and r and thus for the level of consumption.

Several considerations suggest that the typical individual's preferences regarding the utility discount rate will not be reflected in the market.[10] One reason is that, because of externalities, the utility discount rate that most people would endorse as socially conscious beings is lower than the rate of time preference emanating from market transactions. Many years ago, Ramsey, Pigou and Harrod insisted that the only ethically justifiable value for δ is zero. Solow (1974, p. 9) put the matter thus: "In solemn conclave assembled, so to speak, we ought to act as if the social rate of (pure) time preference were zero." A less extreme view would be to say that even if we don't treat δ as zero, individuals do derive a positive externality (outside of the marketplace) from the welfare of future generations. An argument combining the market rate of return on investment with the externality of caring about future generations might call for "low" values of δ, in the range, say, of 0–0.5% per annum.

What about the other term – ηg – in the expression for r? If we again base our choices on a typical individual's preferences, then η should reflect an average person's elasticity of marginal utility of consumption. The value of η is linked to σ, the intertemporal elasticity of substitution in consumption: $\eta = -(\sigma - 1)/\sigma$. Hall's (1988) time series estimates of σ suggest that plausible values for η might lie in the range of 2–4. If we assume per capita growth rates in consumption (g) to have been of the order of 1.5% per annum, we arrive at 3–6% per annum for the term ηg. Together, these considerations would tentatively suggest a value for r in the range 3.0–6.5% per annum.

---

[10] Hence, the justification for a "lower" δ may remain, despite the fact that equilibrium outcomes from individuals' market behavior at first blush suggests a higher δ. On this, see Marglin (1963), Lind (1964) and Sen (1967). In discussing this issue, we have just referred to an externality related to individuals' concerns for future generations' well-being. Elsewhere in this paper, we refer to externalities associated with the use of natural resources. One could interpret individuals' inefficient (excessively rapid) exploitation of natural resources as evidence that they care little about future welfare. However, we prefer the interpretation that such externalities make it impossible for individuals to express their concerns about the future in their individual production or consumption decisions.

How does this value compare with the market rate of return on investment *i*? We need to choose an appropriate market interest rate with which to make the comparison, but rates of return on different securities vary considerably. The real rate of return on private capital in the United States (as directed for use in government project analysis, United States Office of Management and the Budget 1992) is 7.0%; on equities from 1970 to 2000, 7.4%; but on Treasury bills from 1970–2000, just 1.6%. (The last two figures are derived by computation from the Wharton-Data Resources website.) We want to compare the social rate of interest on consumption *r* with a *risk-free* rate. If one interprets Treasury bills as the risk-free asset, then the market rate seems as low or lower than the value for *r* we calculated above. Thus, this rough comparison provides little support for the argument that consumption is excessive.

Two other pieces of evidence that market rates of return on investment may exceed the social rate of interest on consumption come from the incompleteness of markets and the existence of capital taxes. The absence of a complete set of risk-bearing markets, for example, implies that risks cannot be pooled perfectly. Returns from investment are therefore more uncertain, and so recipients (savers) would be expected to attach a lower value to investments than they would have if risks had been pooled more effectively. Consequently, the rate of saving is lower. The absence of complete pooling tends to promote excessive consumption.

Similarly, the taxation of capital income lowers the private return on capital below the social return and thus discourages saving and promotes excessive consumption. However, the normative impacts of capital income taxation depend on other taxes and the prices of various inputs to consumption goods. Labor taxes can discourage labor supply and reduce wage incomes, thus causing current consumption to fall short of the optimal level despite the influence of capital taxes. Moreover, the level of current consumption depends importantly on the prices of the inputs used to produce consumption goods. The pricing of natural resource inputs seems especially important, as we discuss in the following section.

25.2.2 Underpricing of Natural Resources Relative to Social Cost

The level of consumption depends not only on market interest rates, but also on the price of current consumption relative to its social cost or, more specifically, the price of consumption goods relative to capital goods. To the extent that consumption goods are priced below their social cost, consumption will tend to be excessive.

Some natural resources are consumption goods, others are direct or indirect inputs in the production of consumption goods, and many are both. The underpricing of natural resources may contribute to the pricing of consumption goods below their social cost. Such underpricing also alters the relative prices of different consumption goods, thereby leading to inefficiencies in the *composition* (as well as overall level) of consumption: too much consumption of resource-intensive goods and services relative to consumption of other goods. Thus, when natural resource inputs are priced below social cost, both the overall level and the composition of consumption can be affected in ways that lead to excessive natural resource use.

The underpricing of natural resources can stem from at least three sources. First, insecure or poorly defined property rights can lead to excessively rapid resource exploitation if the exploitation does not require much prior investment (Bohn and Deacon 2000). Second, natural resource underpricing can arise from the failure of the market to incorporate the (negative) externalities associated with the use of natural resources. Examples of such externalities include the various damages stemming from the use of fossil fuels (such as acid precipitation or climate change), and the loss of such ecosystem services as flood control, water-filtration and habitat provision when wetlands are drained for conversion to farms.

Third, use of natural resources may be underpriced because of government subsidies. The World Bank's 1992 *World Development Report* (Fig. 3.2) examined fossil fuel, electricity and water prices in 32 developing countries. In all but three of those countries, subsidies caused prices to fall below cost, even before accounting for potential externalities. Similarly, the International Energy Agency (1999) has estimated that in India, China and the Russian Federation, full-cost pricing would reduce energy consumption by 7%, 9% and 16%, respectively. In these countries, most of the departure from social cost pricing is attributed to energy subsidies. For estimates of aggregate global subsidies on the use of environmental and natural resources, see Myers and Kent (2000).

The influence of OPEC on the international market for oil could function as a counterbalance to the above arguments, potentially raising world oil prices up to or beyond social cost. However, there is as yet no clear consensus as to whether current world prices are above or below social cost. In addition, the pricing of oil might well be the exception to a more general pattern of underpriced natural resources.

The underpricing of natural resource inputs can also reduce the prices of *investment* goods. If such underpricing is especially pronounced for investment goods, it could promote a higher ratio of investment to consumption. This does not undo the problem of excessive use of natural resources, which is a more fundamental concern. If the natural resource inputs in the production of investment goods are underpriced, the rates of depletion of natural resources will be too rapid. The rate of accumulation of manufactured capital may or may not exceed the optimal rate, but *genuine* investment – that is, overall investment, inclusive of changes in stocks of natural capital – is likely to be insufficient because of the underpricing of natural resources and the associated excessively rapid depletion of these resources.

25.2.3 Interdependence in Consumption

Interdependence in consumption can also lead to prices of consumption below social cost. Building on Veblen (1899) and Duesenberry (1949), a small but growing body of empirical work (for example, Frank 1985a, b; Ng 1987; Howarth 1996; Schor 1998) suggests that a person's sense of well-being is based not only on a person's own consumption, but also on the person's consumption relative to a "reference group." When others' consumption rises in comparison, an individual could suffer from a loss of well-being because that person's relative consumption now falls. This interdependence in consumption can be viewed as an externality, which can compel individuals to work harder and consume more to keep up with neighbors. It is individually rational behavior, but collectively suboptimal. A formal "growth model" incorporating these ideas has been developed by Cooper et al. (2001).

However, interdependence in consumption does not necessarily imply that people consume excessively. Suppose that the "relative consumption effect" applies not only to current consumption, but also to future consumption (and to leisure, including sleep). Working harder and consuming more today would then improve one's relative current consumption, but worsen one's relative current "consumption" of leisure and also one's relative future consumption! Thus, the bias from interdependence depends on the strength of the effects along various margins. If the effect on individual well-being of relative consumption is symmetric, operating equally on relative consumption and relative leisure now and in the future, it may have no impact on current behavior relative to what would occur if this effect were absent. In this case, the relative consumption effect operates like a lump-sum tax, reducing a person's sense of well-being without changing that person's allocation of labor resources or income.

25.2.4 General Findings

We have identified several factors influencing consumption and have shown how they enable us to judge whether consumption is excessive according to the maximize present value criterion. Several factors – the inability to pool risks perfectly, the taxation of capital income and the underpricing of natural resources – contribute toward excessive consumption. (Below, we will observe that these same factors also work toward excessive consumption according to the sustainability criterion.) Among these imperfections, the underpricing of natural resources strikes us as the most transparent.

25.3 Empirical Evidence Relevant to the Sustainability Criterion

25.3.1 A First Step: Measuring Genuine Investment

As mentioned, genuine investment at date t is the sum of the values of changes in capital stocks at t, evaluated at their accounting prices. It is the change in genuine wealth at constant accounting prices. The assets to be included are manufactured capital, human capital, natural capital and the knowledge base (Dasgupta and Mäler 2000; Arrow et al. 2003b). In this subsection, we begin evaluating whether various nations are meeting the sustainability requirement by estimating and observing the sign of genuine investment. In the next subsection, we will extend the analysis to account for changes in population and technological change.

A growing body of research now aims to measure genuine investment in various countries. The lead has been taken by Kirk Hamilton and his collaborators at the World Bank (for example, Pearce et al. 1996; Hamilton and Clemens 1999; Hamilton 2000, 2002). Hamilton and Clemens (1999), in particular, offered estimates of genuine investment for nearly every nation for the year 1998. They estimated genuine investment by first adding net investment in human capital to existing country-level estimates of investment in manufactured capital. They then made a further adjustment to account for disinvestment in natural resources and environmental capital.[11]

To estimate the accumulation of manufactured capital, Hamilton and Clemens (1999) used figures for net national saving. To estimate the accumulation of human capital, they used expenditure on education. To account for disinvestments in natural resources and environmental capital, they considered the net changes in the stocks of commercial forests, oil and minerals and in the quality of the atmosphere in terms of its carbon dioxide content.[12]

Interestingly, the authors found that in 1998, genuine investment was positive in all the rich nations of the world and in many of the poorer nations as well. However, for 33 of the world's poor countries, including many of the nations of north Africa and sub-Saharan Africa, they estimated that genuine investment was negative.

Here we adopt the Hamilton–Clemens approach as a starting point for evaluating whether selected nations and regions meet the sustainability criterion. Rather than calculate the figures for a single year (as in Hamilton–Clemens), we have calculated annual averages over the past 3 decades, using annual data from the World Development Indicators published in the World Bank's website at ⟨http://devdata.worldbank.org/dataonline⟩. We present our estimates in Table 25.1. The table includes poor nations (China, nations in the Indian subcontinent and nations in the sub-Saharan Africa region), oil-exporting countries (in the Middle East/North Africa region) and industrialized nations (the United States and the United Kingdom).

The differences between genuine investment and the standard measure, net domestic investment, are particularly striking for the Middle East/North Africa and sub-Saharan Africa regions. In these regions, the loss of natural resources more than offset the accumulation of manufactured capital (as reflected in domestic net investment) and human capital (as reflected in expenditure on education). For the

[11] Hamilton and Clemens (1999) employ the term "genuine domestic saving," which we treat as synonymous with genuine investment. The two cannot be distinguished from the data employed in their study, because the data do not include international capital flows.

[12] In calculating the reduction in natural resource stocks, Hamilton and Clemens do not include discoveries of new reserves as an offsetting element. As an accounting matter this seems correct. It is consistent with the notion that the overall global stock of mineral or fuel resources is given – that devoting resources toward exploration, and the subsequent discovery of previously unknown reserves, does not constitute an enlargement of the stock. At the same time, expenditure on exploration equipment or structures should constitute a positive investment in physical capital.

Table 25.1 Genuine investment and components as percentage of GDP

Country	Domestic net investment	Education expenditure	Natural resource depletion: Damage from CO_2 emissions	Energy depletion	Mineral depletion	Net forest depletion	Genuine investment
Bangladesh (1973–2001)	7.89	1.53	0.25	0.61	0.00	1.41	7.14
India (1970–2001)	11.74	3.29	1.17	2.89	0.46	1.05	9.47
Nepal (1970–2001)	14.82	2.65	0.20	0.00	0.30	3.67	13.31
Pakistan (1970–2001)	10.92	2.02	0.75	2.60	0.00	0.84	8.75
China (1982–2001) (without 1994)	30.06	1.96	2.48	6.11	0.50	0.22	22.72
Sub-Saharan Africa (1974–1982; 1986–2001)	3.49	4.78	0.81	7.31	1.71	0.52	–2.09
Middle East & North Africa (1976–1989; 1991–2001)	14.72	4.70	0.80	25.54	0.12	0.06	–7.09
United Kingdom (1971–2001)	3.70	5.21	0.32	1.20	0.00	0.00	7.38
United States (1970–2001)	5.73	5.62	0.42	1.95	0.05	0.00	8.94

Source: Authors' calculations, using data from World Bank (2003).

United States and the United Kingdom, estimated genuine investment exceeded domestic net investment, since the increase in human capital exceeded the value of natural resource depletion.

These figures give an initial glimpse of the change in *V*. We will shortly consider how the picture is altered when changes in technology and in population are taken into account. Before doing so, however, it is useful to consider potential biases in the estimation of genuine investment.[13] First, as Hamilton and Clemens emphasize, one serious problem is the lack of comprehensive data in national accounts. For many important (and declining) natural resource stocks, data are not available – at least not for every country. Among the natural resources not included in the Hamilton–Clemens study are water resources, forests as agents of carbon sequestration, fisheries, air and water pollutants, soil and biodiversity. So there is an undercount, possibly a serious one.

Second, these estimates (as well as earlier estimates by Hamilton and Clemens) employ market prices in indicating potential trade-offs among different forms of capital. To the extent that natural capital is typically underpriced, the use of market prices could bias the estimate of genuine investment in an upward direction.

Third, these estimates consider only very broad categories of capital. At a high level of aggregation, one can miss critical bottlenecks that impose limits on substitution possibilities. The rural poor in the world's poorest countries, for example, often cannot find substitutes when their water holes vanish and the local woodlands recede (Dasgupta 1993). Aggregated calculations can easily miss these details and underestimate the accounting prices of local resource bases.

Fourth, in these calculations, disinvestment in environmental capital is calculated simply as the estimated damage associated with annual emissions of carbon dioxide. A ton of carbon dioxide emission is assumed to lead to a loss of $20 (U.S.) of environmental capital. However, to the extent that newly manufactured capital is expected to yield increased carbon dioxide emissions and thereby inflict subsequent damage, the value of investment in manufactured capital is reduced.[14] The Hamilton–Clemens approach adopted here ignores this effect. A further problem is that this approach implicitly links all of the damage from carbon dioxide emissions to the country responsible for the emissions, whereas in fact a given country's emissions of carbon dioxide produce climate impacts worldwide.

Set against these considerations is another issue that is ambiguous in its effect. The proxy for increases in human capital – expenditure on education – neglects depreciation of human capital due to morbidity, mortality and retirement from the workforce. In this respect, it overstates the increase in human capital. On the other

[13] For a more extensive discussion of ways in which genuine investment could be better estimated, see Arrow et al. (2003b).

[14] This does not double count. In today's genuine investment calculation, today's emissions of CO_2 constitute a reduction in "climate-system capital." This negative investment is the present value of the damages associated with these *current* emissions. At the same time, the value of today's investments in manufactured capital should be the present value of the net services associated with these manufactured assets. Future emissions and climate damages associated with these assets reduce these net service streams and imply a reduction in the value of current investment in these assets.

hand, this proxy ignores skills gained through channels other than formal education, as well as improvements in human productivity that are due to expenditures on health and nutrition. This implies the opposite bias.

While the measure of genuine investment in Table 25.1 conveys useful information, it does not consider changes in population or technology. If population grows fast enough, the long-run productive base per person could decline even if genuine investment were positive. On the other hand, accounting for improvements in productivity would likely brighten the picture. In what follows, we extend our empirical assessment to consider the impacts of population growth and changes in total factor productivity.

25.3.2 Population Growth, Technological Change and Sustainability

The first column in Table 25.2 reproduces the figures from the last column of Table 25.1, which are estimates of average genuine investment as a proportion of GDP over the interval 1970–2001. The second column is an initial, unadjusted estimate of the growth rate of genuine wealth. We arrive at this column by multiplying the numbers in the first column by a presumed GDP/wealth ratio for the country or region in question – the assumed average GDP/wealth ratio over the 3-decade interval. Published estimates of GDP-wealth (or "output/wealth") ratios traditionally have been taken to be something like 0.20–0.30 per year. However, a wide array of capital assets like human capital and many types of natural capital are missing from national accounts. To offset this bias, Table 25.2 uses the figure 0.15 per year for poor and oil-rich countries and regions and 0.20 per year for industrialized countries. Let W refer to per capita genuine wealth. We arrive at the growth rate of W (column 4) by subtracting the population growth rate (column 3) from the growth rate of genuine wealth.

Column 4's figures for changes in per capita genuine wealth do not account for technological change. The remaining adjustments in the table are intended to account for such change as measured through changes in total factor productivity. Column 5 offers estimates of the growth of the total factor productivity residual, as reported for the period 1970–2000 in Klenow and Rodríguez-Clare (1997).[15] For the Middle East/North Africa and Sub-Saharan Africa regions, we obtain the residual by taking a weighted average of the estimates for the countries within each region, using GDP as weights. In the case of China, we report estimates from Collins and Bosworth (1996) for a comparable period, since Klenow and Rodríguez-Clare did not offer estimates for this country. The residual was negative only in the Middle East/North Africa region.

[15] Klenow and Rodríguez-Clare report the growth rates of a transform of total factor productivity. From this information we calculate the associated growth rate of total productivity.

Table 25.2 Growth rates of per capita genuine wealth

	Genuine investment as percent of GDP	Growth rate of unadjusted genuine wealth	Population growth rate	Growth rate of per capita genuine wealth – before TFP adjustment	TFP growth rate	Growth rate of per capita genuine wealth – after TFP adjustment	Growth rate of per capita GDP
Country	(1)	(2)	(3)	(4)	(5)	(6)	(7)
Bangladesh	7.14	1.07	2.16	–1.09	0.81	0.30	1.88
India	9.47	1.42	1.99	–0.57	0.64	0.54	2.96
Nepal	13.31	2.00	2.24	–0.24	0.51	0.63	1.86
Pakistan	8.75	1.31	2.66	–1.35	1.13	0.59	2.21
China	22.72	3.41	1.35	2.06	3.64	8.33	7.77
Sub-Saharan Africa	–2.09	–0.31	2.74	–3.05	0.28	–2.58	–0.01
Middle East/North Africa	–7.09	–1.06	2.37	–3.43	–0.23	–3.82	0.74
United Kingdom	7.38	1.48	0.18	1.30	0.58	2.29	2.19
United States	8.94	1.79	1.07	0.72	0.02	0.75	1.99

Note: These calculations employ the following parameters: output-capital ratio, poor countries/regions 0.15; output-capital ratio, rich countries 0.20; α (share of human and reproducible capital in output) 0.58.

Data for genuine investment, population growth, and GDP growth derive from the World Bank (2003). The genuine investment percentages of GDP derive from data over the time-intervals indicated in Table 25.1. The population growth rate is the average rate over the period 1970–2000.

The estimate for China's total factor productivity (TFP) growth is from Collins and Bosworth (1996). For all other countries or regions, the estimates are from Klenow and Rodríguez-Clare (1997).

We use the numbers in column 5 to arrive at an estimate of the change in per capita genuine wealth that accounts for the impact of projected technological change. Here we employ the approximation formula referred to earlier (and derived in footnote 7). The formula asserts that the contribution of total factor productivity growth to movements in intertemporal social welfare expressed in units of overall capital – that is, the contribution to movements in $V/(p_K K)$ – is γ/α^*, where γ is the growth rate of the total factor productivity residual and α^* is the elasticity of output with respect to all (manufactured, human and natural) capital. α^* can be calculated as the sum of the exponents of each of the types of capital in the aggregate production function. Ideally, one would want to calculate α^* from a production function that includes natural capital. However, we do not have estimates from such a production function. Instead, we rely on the aggregate production function estimated by Klenow and Rodríguez-Clare (1997), which includes human and manufactured capital only. The exponents on these two types of capital are 0.30 and 0.28, respectively. This implies $\alpha^* = 0.58$ and suggests that the adjustment for total factor productivity growth is 1.72 times the value in column 5 of Table 25.2. This is added to column 4 to yield column 6, our adjusted estimate of the growth rate of W (per capita genuine wealth), which takes account of total factor productivity growth.

This adjustment may well overstate the contribution of total factor productivity growth, for two reasons. First, the omission of natural capital biases downward the estimate of α^* and thus biases upward the total factor productivity adjustment, γ/α^*. Suppose that in the true aggregate production function, natural capital's share or exponent were 10%, and the shares of human and reproducible capital were therefore 10% lower than indicated above (to preserve constant returns). In this case, α^* would be 0.62 and the adjustment would be 1.61 (rather than 1.72) times the value in column 5. Moreover, as mentioned earlier, the neglect of natural capital in the production function implies that γ will be misestimated, with the direction of the error depending on the difference in the growth rates of natural resource extraction and overall output. We find that growth rates of energy extraction (a potential proxy for natural resource extraction) tend to exceed GDP growth rates over the past 3 decades, which implies an upward bias in the estimate of γ. Thus, both γ and α^* may be misestimated in a way that leads to an exaggeration of the impact of technological change. In future work, we hope to arrive at improved estimates of the adjustment factor γ/α^*. However, for the present paper we employ in Table 25.2 the values that emerge directly from the Klenow and Rodríguez-Clare (1997) estimates, which appear to be the most satisfactory estimates currently available.

25.3.3 The Poor World

Our (rather tentative) estimates in Table 25.2 suggest that parts of the poor world and some oil exporting regions are not meeting the sustainability criterion. Population growth implies significant differences between the initially estimated growth rate of genuine wealth (column 2) and the growth rate on a per capita basis (column 4). And the results in column 6, which account for both population growth and technological change, differ significantly from the original genuine investment calculations.

Pakistan, Bangladesh, India and Nepal show positive values for the growth rate of *W*. However, estimated growth rates for per capita genuine wealth are projected to be negative in the Sub-Saharan Africa and Middle East/North Africa regions. At an annual rate of decline of 2.6% in genuine wealth per annum, the average person in the sub-Saharan Africa region becomes poorer by a factor of 2 about every 25 years. The ills of sub-Saharan Africa are routine reading in today's newspapers and magazines. But they are not depicted in terms of a decline in wealth.

In the Middle East/North Africa region, per capita genuine wealth is estimated to be declining at an annual rate of 3.8%, after adjusting for technological change. For this region, the estimate for the growth rate in total factor productivity is negative (–0.23%). Hence, adjusting for technological change for this region further increases the estimated decline of per capita genuine wealth. It is worth noting that the negative estimate for total factor productivity growth depends closely on the value one attaches to extracted energy resources. Alternative assumptions about future energy prices can significantly influence this valuation and the estimate of total factor productivity. Thus, the uncertainties in the estimates are especially great in the case of the Middle East/North Africa region.

The results for China are strikingly different from those just discussed. The unadjusted growth rate of per capita genuine wealth is significantly positive – about 2.1%. The estimated growth rate of total factor productivity (3.6%) is high compared with the other countries examined, helping to augment the growth rate of *W*. However, China's figure could be biased upward: the estimates of genuine investment do not include soil erosion or urban pollution, both of which are thought by experts to be especially problematic in China.

How do changes in wealth per head compare with changes in conventional measures of economic progress? The far-right column of Table 25.2 contains estimates of the rate of change of GDP per head over the interval 1970–2000. In all of the listed countries except for China, the growth rate of per capita GDP is considerably higher than the estimated growth rate of per capita genuine wealth. In the Middle East/North Africa Region, per capita GDP growth is positive, while per capita genuine wealth is falling. For all of these countries except China, an assessment of long-term economic development would be significantly off the mark if it were to look at growth rates in GDP per head.

One might infer from the table that several poor countries are "consuming too much." Such a conclusion would be off the mark. In many poor nations, the production of both capital goods and consumption goods is highly inefficient. The countries simultaneously suffer from low levels of genuine investment *and* consumption, and in the most important sense of the term, these nations do not overconsume. One cannot assure a satisfactory quality of life in these nations simply by devoting a larger share of productive factors toward the production of capital goods. Indeed, devoting a greater share of output toward investment could cause considerable misery by reducing what is already a very low level of per capita consumption.

For these nations, the sustainability problem is part of a larger problem of inefficient production and low productivity, which accounts for both low genuine investment and low consumption.

25.3.4 The Rich World

Table 25.2 also displayed estimated growth rates of *W* for the United States and the United Kingdom. These are positive for both countries, although the U.S. estimate is about a third of the UK estimate. The estimates of growth rates of per capita GDP are also positive. Thus, for these countries, the differences across the two indices of economic development are less dramatic.

One might be tempted to conclude that the rich countries are avoiding consuming too much. But it is important to note that the figures for changes in per capita wealth for different countries are not entirely independent. The "success" of rich countries may in part be due to the "failure" of poorer nations. As we noted earlier, natural capital is very frequently underpriced in the market because property rights may be poorly defined or poorly enforced. In extreme cases, such capital assets are free. Dasgupta (1990) and Chichilnisky (1994) have used this fact to argue that countries that are exporting resource-based products (they are often among the poorest) are to an extent subsidizing the consumption of those countries that are importing these products (they are often among the richest). Such hidden subsidies would help promote positive growth rates in per capita wealth in rich countries, while working toward reduced growth rates in the poorer nations that export resource-based products. High levels of consumption in rich countries may promote excessive resource degradation in poor countries, which imperils well-being in the poorer countries. This negative by-product of rich nations' consumption is not captured in existing measures of changes in per capita wealth.

25.3.5 Sensitivity of Results to the Assumed GDP-Wealth Ratio

One significant parameter used in our estimates of the growth of *W* is the GDP-wealth ratio. This parameter is employed to translate genuine investment into a growth rate of (unadjusted) genuine wealth. Table 25.3 indicates the sensitivity of our estimates to this ratio. The middle column reproduces the results in Table 25.2 on the growth rate of per capita annual wealth. The left and right columns, respectively, consider a lower and higher value for the GDP-wealth ratio. For countries with positive genuine investment, a lower value for this ratio implies a higher beginning-of-period level of wealth; thus, for a given level of genuine investment, it implies a lower growth of adjusted per capita genuine wealth. For the regions with negative genuine investment (the Sub-Saharan Africa and Middle East/North Africa regions), a lower value implies a higher (that is, less negative) growth rate of genuine wealth. Results seem fairly

Table 25.3 Sensitivity analysis

Country	GDP-wealth ratio = 0.10	Central case	GDP-wealth ratio = 0.25
Bangladesh	−0.15	0.30	0.93
India	−0.01	0.54	1.41
Nepal	−0.09	0.63	1.91
Pakistan	0.03	0.59	1.34
China	6.79	8.33	10.20
Sub-Saharan Africa	−2.50	−2.58	−2.82
Middle East/North Africa	−3.44	−3.82	−4.51
United Kingdom	1.49	2.29	2.59
United States	−0.15	0.75	1.19

sensitive to this parameter. When the output-wealth ratio is 0.10, the estimated growth of *W* becomes negative in Bangladesh, India, Nepal and the United States.

In the light of the sensitivity of these results to this parameter – as well as significant uncertainties about the underlying data and other parameters that enter these calculations – our conclusions must be very tentative. But despite the uncertainties, it seems clear that measures of changes in per capita genuine wealth yield a very different – and often much bleaker – picture of the prospects for poor nations, as compared with the message implied by changes in GDP per capita.

25.4 Further Perspectives and Conclusions

We have evaluated consumption levels according to two criteria: the discounted present value of the utility stream (the maximize present value criterion) and the maintenance or improvement of intertemporal social welfare (the sustainability criterion). Although the evidence is far from conclusive, we find some support for the view that consumption's share of output is likely to be higher than that which is prescribed by the maximize present value criterion. We also find evidence that several nations of the globe are failing to meet a sustainability criterion: their investments in human and manufactured capital are not sufficient to offset the depletion of natural capital. This investment problem seems most acute in some of the poorest countries of the world.

We would emphasize that insufficient investment by poor countries does not imply excessive consumption in the most important sense. For many of the poorest nations of the world, where productivity and real incomes are low, both consumption and investment are inadequate: current consumption does not yield a decent living standard for the present generation, and current investment does not assure a higher (or even the same) standard for future generations.

This study and all previous studies of which we are aware provide only point estimates of genuine investment or of changes in genuine wealth. Given the vast uncertainties associated with the estimates, even when point estimates are positive, there may remain a significant possibility that genuine investment is negative. The uncertainties justify added caution.

The presence of nonlinearities compounds the importance of uncertainty. The biophysical impacts associated with the loss of natural capital can be highly nonlinear: these impacts may be small over a considerable range, and then become immense once a critical threshold is reached. Crossing the threshold leads to a "bifurcation," a situation where the characteristics of the natural system change fundamentally. For example, shallow clear fresh water lakes can absorb a low level of phosphorus with little ill effect. However, if more phosphorus is added to the lake through sewage or runoff from agricultural land, more algae grows in the water, less sunlight will reach the bottom, and the green plants on the bottom will disappear. As a consequence, the bottom sediments, which contain phosphorus from dead algae, will be less stable, and phosphorous will be released from the bottom. More discharge of phosphorus from outside will trigger even more phosphorus from the bottom. This positive feedback will eventually force the lake to flip or "bifurcate" from clear to turbid water. Recent work by ecologists and environmental economists has investigated these dynamics and shown that such flips can occur over as short a period as a month (Scheffer 1998; Carpenter et al. 1999; Brock and Starrett 2003). Another possible bifurcation arises in certain climate models that indicate that a rise in greenhouse gases might reverse the direction of the Atlantic stream that now warms northern Europe. Paleo-climatic history shows that such reversals have been common.[16]

Nonlinearities in ecosystem dynamics imply the presence of serious downside risks related to the losses of natural capital. Central estimates of the shadow prices for natural capital are likely to be too low if one only considers central cases rather than the entire distribution of potential outcomes from losses of natural capital. Thus, the expected values of genuine investment or of changes in per capita genuine wealth might well be lower than the values emerging when one employs only central values for parameters. Accounting for risk–aversion could lower these estimates even further.

While there may be uncertainty about whether various countries are meeting the sustainability criterion, the need for vigorous public policies to support more efficient consumption and investment choices is unambiguous. Through regulation, taxes or the establishment of clearer or more secure property rights, public policy can help prices of natural and environmental resources better approximate their social cost. These policies can help prevent excessive resource depletion and promote higher genuine investment. Such policies are justified on efficiency grounds whether or not genuine investment currently is positive.[17] Although the

[16] Mastrandrea and Schneider (2001) have employed a climate-economy model to investigate the possibility of reversals and to assess the implications for climate policy.

[17] The rationale for many public policies remains strong irrespective of whether genuine investment is positive or negative. From this, one might conclude that quantifying genuine investment is not very useful. But measuring genuine investment still has significant value. By providing an overall "scorecard" as to whether a nation is investing enough to sustain the welfare of future generations, it can offer an important summary assessment that can help mobilize the general public and politicians.

implementation of public policies that satisfy a benefit–cost test does not in theory guarantee sustainability, such policies do not necessarily conflict with sustainability, either. Indeed, our sense is that policies of this sort – especially those that deal with underpricing of natural resources or environmental amenities – will improve matters along the sustainability dimension. When one views desertification in China, water contamination in Senegal and forest depletion in Haiti, it is hard to imagine that the establishment of property rights or improved pricing of natural resources could worsen the prospects of future generations.

Beyond policy action, further research is needed to identify the areas where current consumption poses a threat to sustainability and to quantify the potential losses. We need to develop better data quantifying the losses of natural capital and the potential for substitution between various forms of capital.

Further, to complement the rather simple analytical calculations of changes in genuine wealth, we need to make more use of disaggregated numerical growth models. Such models can contain considerable detail in the interaction of various forms of capital and the services they generate. Current estimates of genuine wealth depend crucially on the values assigned to shadow prices, yet the empirical basis for these prices is very weak. Numerical growth models can be used to project growth paths of economies and the sensitivity of these paths to changes in capital stocks. In this way, they can generate much better assessments of the critical shadow prices. Additional information of this kind will help reduce uncertainties about changes in genuine wealth and clarify the extent to which current consumption levels might imperil the quality of life of future generations.

Acknowledgements We are grateful to Geir Asheim, Jack Pezzey and the editors of this journal for very helpful suggestions, and to Oren Ahoobim and Justin Diener for excellent research assistance. We also thank the William and Flora Hewlett Foundation for financial support and the Beijer Institute of Ecological Economics, Stockholm, for sponsoring the workshops from which this paper emerged.

References

Arrow, K. J. (2000). Observations on social capital. In P. Dasgupta & I. Serageldin (Eds.), *Social capital: A multifaceted perspective* (pp. 3–5). Washington, DC: World Bank.

Arrow, K. J., Dasgupta, P., & Mäler, K.-G. (2003a). The genuine savings criterion and the value of population. *Economic Theory, 21*(2), 217–555.

Arrow, K. J., Dasgupta, P., & Mäler, K.-G. (2003b). Evaluating projects and assessing sustainable development in imperfect economies. *Environmental and Resource Economics, 46*(4), 647–685.

Arrow, K. J., & Kurz, M. (1970). *Public investment, the rate of return and optimal fiscal policy.* Baltimore, MD: Johns Hopkins University Press.

Arrow, K. J., et al. (1996). Intertemporal equity, discounting, and economic efficiency. In J. P. Bruce, H. Lee, & E. R. Haites (Eds.), *Climate change 1995: Economic and social dimensions of climate change* (pp. 125–44). Cambridge: Cambridge University Press.

Asheim, G. (1994). Net national product as an indicator of sustainability. *Scandinavian Journal of Economics, 96*(2), 257–265.

Asheim, G. (2003). Green national accounts for welfare and sustainability: A taxonomy of assumptions and results. *Scottish Journal of Political Economy, 50*(2), 113–130.

Barnett, H. J., & Morse, C. (1963). *Scarcity and growth*. Baltimore, MD: Johns Hopkins University for Resources for the Future.

Bohn, H., & Deacon, R. T. (2000). Ownership risk, investment, and the use of natural resources. *American Economic Review, 9*(3), 526–549.

Brock, W. A., & Starrett, D. (2003). Managing systems with non-convex positive feedbacks. *Environmental and Resource Economics, 46*(4), 575–602.

Carpenter, S. R., Ludwig, D., & Brock, W. A. (1999). Management of eutrophication for lakes subject to potentially irreversible change. *Ecological Applications, 9*(3), 751–771.

Chichilnisky, G. (1994). North-south trade and the global environment. *American Economic Review, 84*(3), 851–874.

Collins, S., & Bosworth, B. (1996). Economic growth in East Asia: Accumulation versus assimilation. *Brookings Papers on Economic Activity, 2*, 135–203.

Cooper, B., Garcia-Peñalosa, C., & Funk, P. (2001). Status effects and negative utility growth. *Economic Journal, 111*(2), 642–65.

Daily, G. C. (Ed.) (1997). *Nature's services: Society's dependence on natural ecosystems*. Washington, DC: Island Press.

Daily, G. C., et al. (2001). The value of nature and the nature of value. *Science, 289*, 395–396.

Dasgupta, P. (1990). The environment as a commodity. *Oxford Review of Economic Policy, 6*(2), 51–67.

Dasgupta, P. (1993). *An inquiry into well-being and destitution*. Oxford: Clarendon Press.

Dasgupta, P. (2001a). *Human well-being and the natural environment*. Oxford: Oxford University Press.

Dasgupta, P. (2001b). On population and resources: Reply. *Population and Development Review, 27*(4), 748–54.

Dasgupta, P., & Heal, G. M. (1979). *Economic theory and exhaustible resources*. Cambridge: Cambridge University Press.

Dasgupta, P., & Mäler, K.-G. (2000). Net national product, wealth and social well-being. *Environment and Development Economics, 5*, 69–93.

Duesenberry, J. S. (1949). *Income, saving, and the theory of consumer behavior*. Cambridge, MA: Harvard University Press.

Ehrlich, P. R., & Ehrlich, A. H. (1990). *The population explosion*. New York: Simon and Schuster.

Frank, R. (1985a). *Choosing the right pond: Human behavior and the quest for status*. New York: Oxford University Press.

Frank, R. (1985b). The demand for unobservable and other nonpositional goods. *American Economic Review, 75*(1), 101–16.

Hall, R. (1988). Intertemporal substitution in consumption. *Journal of Political Economy, 96*(2), 339–57.

Hamilton, K. (2000). Genuine saving as a sustainability indicator. Working paper, Environment Department, World Bank, January.

Hamilton, K. (2002). Savings rules and Sustainability. Working paper, Environment Department, World Bank.

Hamilton, K., & Clemens, M. (1999). Genuine savings rates in developing countries. *World Bank Economic Review, 13*(2), 333–56.

Heal, G. M. (1998). *Valuing the future: Economic theory and sustainability*. New York: Columbia University Press.

Heal, G. M. (2003). Optimality or sustainability? In R. Arnott, B. Greenwald, R. Kanbur, & B. Nalebuff (Eds.), *Economics for an imperfect world* (pp. 331–48). Cambridge, MA: MIT Press.

Heal, G. M., et al. (2001). Protecting natural capital: Ecosystem service districts. *Stanford Environmental Law Journal*, 20(2), 333–64.

Howarth, R. (1996). Status effects and environmental externalities. *Ecological Economics, 16*(1), 25–34.

Hussain, A., Stern, N., & Stiglitz, J. (2000). Chinese reforms from a comparative perspective. In P. J. Hammond & G. D. Myles (Eds.), *Incentives, organization, and public economics* (pp. 243–77). Oxford: Oxford University Press.

International Energy Agency. (1999). *World energy outlook – 1999 insights. Looking at energy subsidies: Getting the price right*. London: IEA Publications.

Johnson, D. G. (2000). Population, food, and knowledge. *American Economic Review, 90*(1), 1–14.

Johnson, D. G. (2001). On population and resources: A comment. *Population and Development Review, 27*(4), 739–47

Klenow, P. J., & Rodríguez-Clare, A. (1997). The neoclassical revival in growth economics: Has it gone too far? In B. Bernanke & J. Rotemberg (Eds.), *NBER Macroeconomics Annual 1997* (pp. 73–104). Cambridge, MA: MIT Press.

Koopmans, T. C. (1960). Stationary ordinal utility and impatience. *Econometrica, 28*(2), 287–309.

Koopmans, T. C. (1972). Representation of preference orderings over time. In C. B. McGuire & R. Radner (Eds.) *Decision and organization* Chap. 4. North Holland: Amsterdam.

Levin, S. (Ed.). (2001). *Encyclopedia of biodiversity*. New York: Academic Press.

Lind, R. C. (1964). The social rate of discount and the optimal rate of investment, MD: Further comment. *Quarterly Journal of Economics, 78*(2), 336–45.

Lind, R. C. (Ed.) (1982). *Discounting for time and risk in energy policy*. Baltimore: Johns Hopkins University Press.

Marglin, S. A. (1963). The social rate of discount and the optimal rate of investment. *Quarterly Journal of Economics, 77*(1), 95–111.

Mastrandrea, M., & Schneider, S. H. (2001). Integrated assessment of abrupt climatic changes. *Climate Policy, 1*(4), 433–49.

McNeill, J. R. (2000). *Something new under the sun: An environmental history of the twentieth-century world*. New York: W.W. Norton.

Myers, N., & Kent, J. (2000). *Perverse subsidies: How tax dollars can undercut the environment and the economy*. Washington, DC: Island Press.

Ng, Y.-K. (1987). Relative income effects and the appropriate level of public expenditure. *Oxford Economic Papers, 39*(2), 293–300.

Pearce, D., & Atkinson, G. (1995). Measuring sustainable development. In D. W. Bromley (Ed.), *Handbook of environmental economics* (pp. 166–81). Oxford: Basil Blackwell

Pearce, D., Hamilton, K., & Atkinson, G. (1996). Measuring sustainable development: Progress on indicators. *Environment and Development Economics, 1*, 85–101.

Pezzey, J. (1992). Sustainable development concepts: An economic analysis. World Bank Environment Paper No. 2. Washington, DC: World Bank.

Portney, P., & Weyant, J. (Eds.). (1999). *Discounting and intergenerational equity*. Washington, DC: Resources for the Future.

Postel, S. L., Daily, G. C., & Ehrlich, P. R. (1996). Human appropriation of renewable fresh water. *Science, 271*, 785–788.

Ramsey, F. P. (1928). A mathematical theory of saving. *Economic Journal, 38*(4), 543–59.

Scheffer, M. (1998). *The ecology of shallow lakes*. London: Chapman & Hall.

Schor, J. (1998). *The overspent American*. New York: Basic Books.

Sen, A. K. (1967). Isolation, assurance, and the social rate of discount. *Quarterly Journal of Economics, 81*(1), 12–124.

Solow, R. M. (1974). The economics of resources or the resources of economics. *American Economic Review, 64*(2), 1–14.

Solow, R. M. (1992). *An almost practical step toward sustainability. Resources for the future 40th anniversary lecture*. Washington, DC: Resources for the Future.

UNDP (United Nations Development Programme). (2003). *Human development report*. New York: Oxford University Press.

United States Office of Management and the Budget. (1992). *Circular A-94 revised*. Washington, DC: United States Government Printing Office.

Veblen, T. (1899). *The theory of the leisure class*. New York: Penguin

Vitousek, P., Ehrlich, P. R., Ehrlich, A. H., & Matson, P. (1986). Human appropriation of the product of photosynthesis. *BioScience, 36*, 368–73.

Vitousek, P. M., Mooney, H. A., Lubchenco, J., & Melillo, J. M. (1997). Human domination of earth's ecosystem. *Science, 277*, 494–499.
WCED. (1987). *Our common future. Report of the world commission on environment and development*. Oxford: Oxford University Press.
World Bank. (1992). *World Bank development report 1992: Development and the environment*. Washington, DC: World Bank.
World Bank. (1997). *Expanding the measure of wealth: Indicators of environmentally sustainable development*. Washington, DC: World Bank.
World Bank. (2003). *World development indicators*. Available at http://devdata.worldbank.org/dataonline, September.

Chapter 26
Askö in Stanford 2000: Commentary by Terry Hughes

Terry Hughes

The global human footprint depends on both population size and how each person impacts on their environment. From 2.5 billion in 1950, the world's population has grown to nearly seven billion people today, and it will almost certainly exceed nine billion by 2050. At the same time, per capita consumption has grown even more rapidly, due to increased affluence and greater access to new technologies and global markets. The environmental impacts of human societies, and the dependence of people on ecosystem services, are undeniable. So from an ecological perspective, the answer to the question "are we consuming too much?" is obviously yes. Human development has come at a significant environmental cost, including the ongoing and growing impacts of pollution, habitat loss, soil degradation, loss of biodiversity, over-fishing, and climate change.

Yet, by most metrics, people are undeniably better off today than at any time in our history. The question posed by Arrow et al., is whether the projected use of earth's resources is endangering future economic options for humanity. Their approach, based on a collaboration of leading economists and ecologists fostered by the Beijer Institute, is to value and account for natural capital in order to simultaneously conserve resources and improve human well-being. Arrow et al. point out that natural resources are usually greatly underpriced. For example, Standard National Accounts generally ignore the value of ecosystem services, although they are unlikely to do so in the future. There is a growing push around the world to assign property rights and economic values to ecosystem services that are traditionally viewed as being free to all-comers. In particular, paying for access to water and carbon sinks is increasingly prevalent in emerging approaches to curbing unsustainable exploitation of natural resources.

A critical issue for evaluation of consumption into the future is the assumption of a constant discount rate, which weights the present utility in comparison to the

T. Hughes
ARC Centre of Excellence for Coral Reef Studies, James Cook University, Australia
e-mail: terry.hughes@jcu.edu.au

T. Söderqvist et al. (eds.), *Bringing Ecologists and Economists Together*,
DOI 10.1007/978-90-481-9476-6_26, © Springer Science+Business Media B.V. 2011

future. Higher discount rates leave fewer resources for future generations. Uncertainty in discount rates, the finite extent of non-renewable resources such as oil and minerals, low rates of regeneration for many exploited species and ecosystems, and the difficulty of measuring how much is left, all argue for a cautionary approach that places limitations on rates of exploitation.

Arrow et al. recognize too that threshold behavior in ecological systems also creates uncertainty in determining safe levels of exploitation. For example, human activities affect coastal coral reefs on three broad fronts – through land use in adjoining catchments, by over-fishing, and by human-induced climate change. Already, many corals are suffering unprecedented high rates of mortality from thermal stress caused by global warming. These three drivers of change are highly interactive with each other. Runoff of sediment and nutrients promotes blooms of seaweed, as does overfishing of herbivores, eroding the resilience of reefs to absorb recurring bouts of mortality from natural and unnatural disasters. Instead of recovering as they have for millennia, many reefs are now degraded, with a reduced capacity to provide income for local societies from fisheries and tourism. The impacts of loss of natural capital on reef ecosystems (e.g. due to land clearing and fishing) are small initially, but become catastrophic at higher levels. Similar, non-linear effects occur with eutrophication of lakes, salinization of soils, and many other biophysical systems that support human well-being. Identifying safe levels of use for human development, and avoiding dangerous tipping-points is crucial for our future prosperity.

Depletion of natural assets is often masked by substitution of one for another, so that incomes are maintained while the environmental impacts extend to different resource or another place. The substitution of local firewood for coal and oil, and the global sourcing and importation of seafood, are familiar examples in the economic and ecological trajectories of the developed world. Poorer countries have fewer options for substitution when local resources are depleted. A key aspect of Arrow et al.'s paper is the issue of inequality in economic development. For example, the impacts of climate change (e.g. from extreme weather events, disease, crop failure) are greatest for low-income countries and regions, whereas most of the emissions to date have been produced by wealthy, industrialized countries. Furthermore, an iniquitous subsidy arises from the under-pricing of natural resources that are exported from poor to rich countries. These hidden subsidies promote positive economic growth in the developed world, at the expense of poorer nations. For the poorest countries in particular, Arrow et al.'s criteria for economic sustainability are not met: The depletion of natural capital exceeds current investments in human and manufactured capital.

Arrow et al.'s paper exemplifies the interface of economic and ecological thinking that characterizes the Askö meetings. Ecologists, economists and social scientists must play a bigger role in informing public debate by explaining the linkages between people and ecosystems, shaping societal support for the equitable and long-term use of ecosystems, and sustaining the incalculable aesthetic and spiritual values of our natural resources.

Chapter 27
Askö in Stanford 2000: Commentary by Thomas Sterner

Thomas Sterner

This is an exciting and important article to comment on: It goes straight to the heart of one of the most vital questions in environmental economics – in fact one of the most vital of all questions for humanity. It is written by a group of most prominent economists and ecologists. The article deftly takes the reader through an analysis of what sustainable consumption means – primarily to the economist. It distinguishes nicely between two criteria: net investments being sufficient to maintain future wellbeing at the current level and maximization of net present consumption. The authors are pedagogical in explaining how these two criteria are related and how they can be extended to consider complications such as technical progress and population growth. They also present some empirical applications based on the genuine savings data generated by the World Bank.

These data paradoxically appear to show that rich countries are sustainable but not poor ones and the authors discuss the obviously unsatisfactory conclusion that it looks as if the poor are consuming too much but not the rich. This is clearly provocative in many respects and I think this is where the paper becomes most interesting and where perhaps more thought might be warranted. For most people the question posed in the title of the paper "Are we consuming too much?" appears primarily a moral question but as the authors show it can also be given a precise technical, economic interpretation. They focus very much on the aggregate and I think it might be constructive to add emphasis to the sectoral and geographical dimensions.

From the viewpoint of ecology – or of a local farmer for that matter – I think the question of aggregate net genuine savings is very hard to interpret. The real constraints are related to water, nutrients, carbon dioxide, CFCs and for each of these the effects can really only be interpreted at the local level. Water stress and shortage are clearly "global problems" in some sense but it is hard to say whether total consumption of water in the world is too high. It is more relevant to discuss such an issue at the level

T. Sterner
Department of Economics, University of Gothenburg, Sweden
e-mail: thomas.sterner@economics.gu.se

T. Söderqvist et al. (eds.), *Bringing Ecologists and Economists Together*,
DOI 10.1007/978-90-481-9476-6_27,

of water catchments. If we waste some water during the monsoon in Bangladesh – or even in Sweden – it does not matter so much as in a dry area. Saving water in say Stockholm does not readily compensate for overuse in the Sahel.

Naturally attention to spatial and ecological diversity can be combined with a holistic economic approach but we then need considerable information on current and future relative prices for many ecosystem services defined in considerable detail. As shown in the article, discounting is crucial to the definition of sustainability and in particular the choice of parameters related to the Ramsey equation. I believe that if we are interested in the sustainability of future growth we need to pay more attention to the characteristics of that growth. Sustainability hinges on *which* sectors, technologies and consumption types are growing and *where*. Growth in computer games is probably sustainable whereas growth in the use of fossil fuel and carbon emissions is not. This would suggest that growth (in the aggregate sense) is possible as long as the growth occurs in some sectors rather than others. Changing sectoral composition in future growth will imply that relative prices will change considerably and we need to integrate our analysis of changing relative prices into the analysis of discounting and that will affect such calculations as net savings or maximized discounted welfare (see Hoel and Sterner 2007; Sterner and Persson 2008).

These issues become particularly poignant when it comes to natural resources in developing countries that are often underpriced or in fact taken as free. The authors do discuss this issue very well but fairly briefly. This is again a very important topic that should be developed more. According to a recent newspaper article a single cup of coffee requires 140 L of water for its production. Irrespective of whether the figure is exact, it is well known that such products as coffee are very water intensive. Naturally it is unlikely that this water is fully priced and it is easy to imagine that some farmers or neighbors suffer because too much water is abstracted so that Kenyan or Brazilian producers can make money selling coffee to European consumers. The shadow cost of this water will vary greatly from one location to another and thus ideally these varying shadow prices should be factored in – not only into the calculations of genuine net savings but into real prices so that socially non-profitable transactions are avoided and so that the consumers have to bear the full cost of their natural resource use while local owners or users of ecosystems should be correspondingly compensated.

We all know that such a World is distant. In the meantime this brings us back to the moral dimension. Should we, the consumers of say coffee, in the rich World do something about our coffee consumption (and meat and sugar and a whole array of other products)? If free abstraction of scarce resources (of a common pool nature) is used to implicitly subsidize our consumption in the rich World then maybe the relevant question is: what are we over-consuming? This raises the issue of whether it is really a good idea to focus on the aggregate. The question should not be whether rich or poor people are consuming too much in the aggregate. The important questions concern individual goods or sectors or ecosystem services – often in specific locations. Are we consuming too much water in the Sahel, are we consuming too much fossil energy? If all our constraints but one were met – but the one not met is sufficiently vital or binding then we will still face dire problems. Which products – from which countries or producers – should we refrain from in order to reduce unsustainability that will be manifested not in our own countries or ecosystems but in those from where the products come? If local governments in producing countries fail (or do not want) to implement effective policies then should we do so in their place? Possible instruments might for instance include labeling schemes or trade instruments but it is also easy to see that there can be serious pitfalls with policies that imply trade barriers.

A final and related issue the paper touches on is interdependence in consumption. Suppose people do not really derive their welfare from just consumption but rather from relative consumption. Suppose in the extreme that most efforts to promote growth in sectors that lead to ecological problems actually derive from a desire to catch up with consumption patterns in the rich World: Does this add to our moral obligations to modify our own consumption patterns? I believe that the paper has given us not only interesting answers and tools for analysis but also that – like every really seminal piece – that it raises further questions to be analyzed. These questions relate to the interplay between ecological and geographical detail, the patterns rather than the rate of growth and the relative accounting prices entailed by differential growth and finally the moral-ethical issues related to how to correct for erroneous prices in other jurisdictions.

References

Hoel, M., & Sterner, T. (2007). Discounting and relative prices. *Climatic Change, 84*, 265–280.

Sterner, T., & Persson, M. (2008). An even sterner report: Introducing relative prices into the discounting debate. *Review of Environmental Economics and Policy, 2*(1), 61–76.

Chapter 28
Askö 2000: Genetic Diversity and Interdependent Crop Choices in Agriculture*

Geoffrey Heal, Brian Walker, Simon Levin, Kenneth Arrow, Partha Dasgupta, Gretchen Daily, Paul Ehrlich, Karl-Göran Mäler, Nils Kautsky, Jane Lubchenco, Steve Schneider, and David Starrett

Abstract The extent of genetic diversity in food crops is important as it affects the risk of attack by pathogens. A drop in diversity increases this risk. Farmers may not take this into account when making crop choices, leading to what from a social perspective is an inadequate level of diversity.

---

*This article was published in *Resource and Energy Economics*, Vol. 26, Heal et al., Genetic diversity and interdependent crop choices in agriculture, 175–184, Copyright Elsevier (2004).

G. Heal
Columbia Business School, Columbia University, Uris Hall, 3022 Broadway, New York, NY 10027-6902, USA

B. Walker
CSIRO, Canberra, Australia

S. Levin
Department of Ecology & Evolutionary Biology, Princeton University, Princeton, NJ, USA

K. Arrow and D. Starrett
Department of Economics, Stanford University, Stanford, CA, USA

P. Dasgupta
Faculty of Economics, Cambridge University, Cambridge, UK

G. Daily, P. Ehrlich, and S. Schneider
Department of Biological Sciences, Stanford University, Stanford, CA, USA

K.-G. Mäler
Beijer Institute of Ecological Economics, the Royal Swedish Academy of Sciences, Box 50005, SE-10405 Stockholm, Sweden

N. Kautsky
Department of Systems Ecology, Stockholm University, Stockholm, Sweden

J. Lubchenco
Department of Zoology, Oregon State University, Corvallis, OR, USA

T. Söderqvist et al. (eds.), *Bringing Ecologists and Economists Together*, DOI 10.1007/978-90-481-9476-6_28, © Springer Science+Business Media B.V. 2011

28.1 Trends in Genetic Diversity in Agriculture

Are we deploying sufficient genetic diversity in agriculture? The level of genetic diversity that characterizes commercially important crops and animals is a matter of considerable concern, as it is generally agreed that access to genetic diversity has been and remains important in maintaining and increasing agricultural productivity. There are economic and epidemiological reasons to believe that the varietal choices made by farmers may lead to less genetic diversity in agriculture than is desirable from a social perspective, implying that policy interventions may be needed to ensure that we choose an appropriate level of diversity. Analyzing why we may systematically choose too little diversity is the main theme of this paper. However, before exploring these issues we review briefly the evidence on what is happening to genetic diversity in agriculture.

Some commentators have suggested that we may have experienced a decline in genetic diversity of commercial crops and animals. They attribute this to the destruction of the native habitats of many breeds and varieties, to the domestication and ensuing development of genetically uniform varieties and breeds, and to farmer and consumer preferences for certain breeds or varieties.[1] Another contributory factor is said to be the consolidation of the seed grain industry globally, leading to a more limited choice of seed varieties. At the same time, there has been widespread adoption of the same high-yield varieties of common food crops in many countries, as a part of the green revolution. (Porceddu et al. 1988) As an illustration, China currently possesses over 50 unique pig breeds, but many of these are becoming endangered as they are replaced with Western breeds. Porceddu et al. (1988) identify three waves of genetic erosion, the first beginning in the nineteenth century in Europe with the advent of the early plant breeders, the second in the mid-twentieth century, associated with a massive effort to improve productivity in non-industrialized countries through breeding for varieties with short stature and disease resistance, and a third associated with the push for uniformity (for example, using "shuttle breeding" to endow high yielding varieties with photoinsensitivity, allowing them to be grown in widely different parts of the world). The most successful examples of such centralized breeding institutes are Centro Internacional de Mejoramiento de Maiz y Trigo (CIMMYT) in Mexico and International Rice Research Institute (IRRI) in the Philippines (Marshall 1977).

Further evidence for this suggestion is provided by Hammer et al. (1996), who analyzed the differences between collecting missions in Albania (1941 and 1993) and in south Italy (1950 and the late 1980s) and claimed high losses in genetic variability: levels of genetic erosion of 72.4% and 72.8%, respectively. In India, rice varieties have declined from an estimated 400,000 before colonialism to 30,000 in the mid-nineteenth century with several thousand more lost after the green revolution in the 1960s; also Greece is estimated to have lost 95% of its broad genetic stock of traditional wheat varieties after being encouraged to replace local seeds

[1] See Committee on Agriculture Science and Technology (2003).

with modern varieties developed by CIMMYT (Lopez 1994). (Lopez also quotes a boast by Stalin to Churchill: "We have improved beyond measure the quality of our wheat. We used to sow all varieties, but now we only cultivate the Soviet prototype. Any other cultivation than that is prohibited nation-wide.")

A rather different perspective is provided by Smale (1997, 1998a,b) and Wood and Lenne (1997). Smale studies the trends in bread wheat varieties in use since the start of the green revolution. She tests two propositions – one that the green revolution caused genetic erosion and the second that it caused an increase in genetic vulnerability. These are significant parts of the cases developed in the studies cited in the previous paragraphs. Smale notes that a reduction in the number of varieties in use does not necessarily imply a reduction in genetic diversity, as there is no clear mapping from a reduction in the number of varieties to a loss of genetic material: indeed, she observes, some new varieties may be crosses that bring new genetic material into the commercial population even in the face of a drop in a count of varieties. She also cites evidence that over the twentieth century the dominance of commercial bread wheats by a small number of varieties decreased, and that some of the land races no longer cultivated are still maintained in seed banks, and indeed have contributed genetic material to currently-used varieties.

Smale also criticizes the argument that a reduction in the number of varieties is increasing vulnerability to disease, an important part of the arguments of those who argue that we should be concerned about loss of genetic diversity. She cites evidence that modern varieties of wheat are less vulnerable to most traditional wheat diseases than their predecessors – this being one of the reasons why they have been adopted widely by farmers. Smale also observes that modern plant breeders aim for polygenic resistance to pathogens, which is intrinsically more durable and stable than the monogenic resistance that characterized early plant breeding. The loss of diversity associated with the introduction of the most resistant genotypes does, however, carry with it the potential for loss of resiliency at the level of the system as a whole.

The US National Research Council (1993) looking into this issue, concluded that the pattern is complex: various components of genetic vulnerability are going up, and others down, simultaneously in different parts of the world and for different crops. Overall, the Committee concludes (p. 80): "For major crops in developed countries, varietal turnover and the number of varieties planted have increased since 1972, indicating that the overall level of genetic vulnerability may have decreased. The genetic basis of elite germplasm, however, was found to be shallow because of extensively shared ancestry and limited use of exotic germplasm." The summary goes on to specify priorities for reducing genetic vulnerability.

28.2 The Role of Genetic Diversity

What is at stake in this discussion? The key point is that a loss of genetic diversity may lead to significant risks for food supplies. A pathogen that attacks the predominant commercial variety of a food crop can inflict immense costs on society. The classic example of this is the Irish potato famine of the nineteenth century. More recently, the loss of a significant fraction of the Asian rice crop to the grassy stunt virus[2] illustrates the same point – the extreme vulnerability of a geographically extensive and genetically homogeneous crop to damage by a well-adapted predator. Once a pathogen attacks any part of such a crop, it can spread rapidly and extensively through a host to which it is well adapted with no natural barriers. In 1970 in the USA, the southern corn leaf blight epidemic resulted in enormous losses; this was considered to be a man-made epidemic caused by excessive homogeneity of the USA's tremendous maize hectarage (Browning 1988). Other examples include the coffee rust epidemic in Ceylon in the 1870s, the tropical maize rust epidemic in Africa in the 1950s and the blue mould epidemic on tobacco in the USA and Europe in the 1960s (Marshall 1977).

International trade can enhance these risks. The rapid movement of cargoes, animals and plants internationally means that once a pathogen occurs anywhere it can travel fast and poses a threat to populations far distant. The spread of foot and mouth disease in Europe illustrates the rapid international spread of a pathogen in the face of massive attempts to prevent this. The spread of the glassy sharpshooter to California is another example.

Not only does the genetic homogeneity of populations raise the cost of a pathogen outbreak should it occur, but it also increases the probability of such an outbreak in the first place. The distribution of pathogens by type is not exogenously given but is influenced by the distribution of host types. Pathogens evolve and adapt to the available hosts. Increasing the scale of a host will increase the probability of a pathogen evolving that is well suited to that host. We can think of the expected loss (risk) from a pathogen epidemic as a function of two variables, the probability of an epidemic and the likely cost of an epidemic given that one occurs. Formally, $L = P(E)C(E)$, where L is the expected loss from an epidemic, $P(E)$ the probability of an epidemic and $C(E)$ the cost of an epidemic. Both magnitudes on the right of this equation are increased by genetic homogeneity – this increases the probability of an epidemic (by increasing the selective pressure on the pathogen to overcome host resistance) and also the cost of one should it occur (by facilitating rapid spread). L is, therefore, likely to increase non-linearly in the degree of homogeneity. Furthermore, genetic homogeneity also serves to increase $P(E)$ by increasing the susceptible pool for a pathogen that can overcome resistance, thereby increasing the rate of spread. Indeed, epidemiological models typically imply that there is a threshold susceptible population size, below which a pathogen cannot spread at all; introducing diversity, therefore, could reduce the susceptible pool for any pathogen low enough that no outbreak is possible, or at least reduce the probability. The simplest example represents the rate of change of the infectious population I as $dI/dt = bSI - kI$, where S is the number of susceptible plants, bSI the rate of new infections, and kI is the death rate of infected plants. If $S < kI$, any introduction of a novel pathogen will die out monotonically.

[2] Myers (1997).

28.3 Genetic Diversity and Risk Management

In growing food, as in many other areas, society faces a risk–return trade off. It can enhance the productivity and average yield of food crops at the cost of greater risk, measured by a higher standard deviation of their yields. One illustration of this is the use of crops well adapted to particular weather patterns. If a suitable weather pattern materializes these can give yields greatly in excess of more generic crops, but if the weather pattern that is realized is not that to which they are finely tuned then their yield may be much lower than that of other options. The farmer faces a higher peak yield, and a higher yield in the normal weather conditions for his region, in exchange for poorer outcomes in less normal weather patterns.

Striking the right balance between risk and return is important: given that many people live at or near subsistence, the consequence of crop failure can be catastrophic. With little or no slack in food supplies in poor countries, society has to watch carefully the risk of a major disruption. At the same time, there is a risk in not pursuing every opportunity to expand food supplies, the risk that the population will outgrow food supplies in some regions. The risks associated with genetic uniformity have been recognized for over half a century and have given rise to attempts to conserve crop diversity, both in situ and ex situ, with arguments in favor of both. In Bush (2000), case studies of crop diversity illustrate that present levels of diversity are the outcome of farmers' needs and preferences, on the one hand, and the increasing availability of a limited number of high yielding varieties on the other – often encouraged by government. Bellon (1996), in an analysis of maize landraces in southern Mexico, found that even though new high yielding varieties and commercial support were available, farmers nevertheless maintained a complex population of landraces because no single variety could satisfy their five main concerns; environmental heterogeneity effects, resistance to pests and pathogens, risk management, culture and ritual, and diet. Browning (1988), in a review of the use of diversity, concludes that diversity is the only defense against the unknown.

There are reasons to think that farmers may be making the wrong risk–yield trade off from the perspective of society in reducing genetic diversity in order to raise yields, and specifically that they are choosing too much risk. The overall level of genetic diversity in agriculture is the outcome of millions of varietal choices made by individual farmers across the world. If they all choose similarly then society as a whole has little genetic diversity in its crops. Each farmer, when choosing a crop or animal variety, will seek what for him is the most appropriate balance between higher average return and greater risk, indicated by greater variability of the yield (see Fig. 28.1). Farmers will always want to choose crop combinations on the risk–yield frontier, and exactly where on this they choose will depend on their tolerance for risk. By way of illustration, those with reserves to fall back on, or those with crop insurance, will be willing to use more risky combinations than those who are liable to starve if their crops fail. When making these choices, farmers will not in general consider the implications of their choices for the overall pattern of diversity and the implications that this has for the risks that society as a

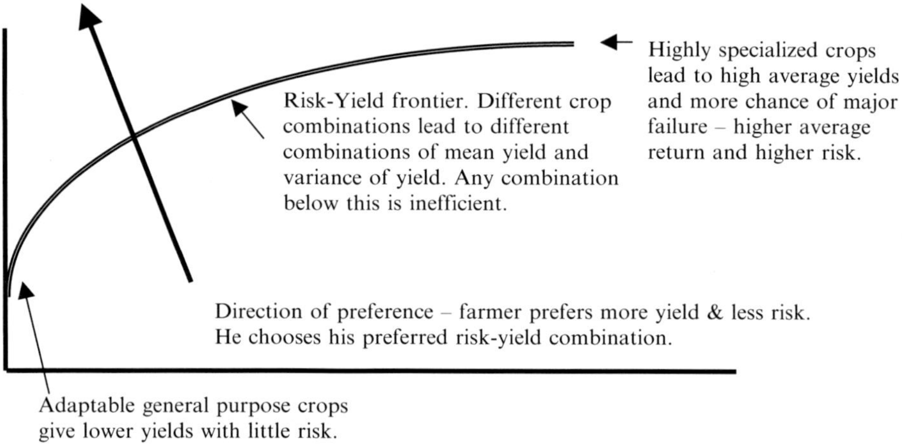

Fig. 28.1 The risk–yield frontier

whole faces. Nor will a farmer necessarily assess accurately the risks that he personally will face as a result of his choices.

The key point here is that the risk that each farmer faces as a result of his crop choice is not fully under his control. Compare two hypothetical situations, one where a farmer plants his 100 ha with crop variety *X* and no one else within 500 km plants variety *X*, and a second one where every farmer within 500 km also plants variety *X*. Our hypothetical farmer faces a greater risk of pathogen damage to his crops in the second case than in the first. If a pathogen for variety *X* arrives anywhere within 500 km, there is a higher probability that it will spread to him in the second case than in the first, as it will find hosts and thrive anywhere within 500 km. In addition, with an area of at least 3.142×500^2 km^2 planted with variety *X*, there is now a chance that somewhere in this region there will occur a mutation of an existing organism that produces a pathogen for variety *X*. This chance, as we have noted, is larger, the greater the area planted with variety *X*.

We can model these issues using a simple game-theoretic framework. Assume for simplicity that there are two farms, F_1 and F_2, and two crop varieties V_1 and V_2. Let p_i be the probability of a severely damaging infestation of crop V_i ($i = 1$ or 2) when it is planted by only one farmer and let q be the probability of an infestation crossing from one farm to the other if both use the same variety (for simplicity, we take q to be independent of the common variety). If the crops are different there is no risk of an infestation moving from one farm to another. If a crop is successful then a farmer's income is Y (independent of the crop again, mainly to simplify notation) and if there is an infestation this is reduced by L, the loss from the infestation. There are obviously four possibilities that we need to consider, represented in the following payoff matrix where the rows are F_1's strategy and the columns F_2's strategy. The cases are: both farms use V_1, both use V_2, and each uses a different variety (V_1/V_2 and V_2/V_1). If each uses a different variety then the expected payoffs in the two cases are $(Y - p_1L, Y - p_2L)$ and $(Y - p_2L, Y - p_1L)$ where the first entry is farm 1's return and the second farm 2's return. (These are the lower left and upper right entries in the payoff matrix.) If both farms plant V_1 (the top left corner in the payoff matrix) then each has the same payoff, $Y - p_1L - (1 - p_1)p_1qL$. Conversely, if both plant V_2 then their common payoff is $Y - p_2L - (1 - p_2)p_2qL$ (bottom right). In the case when both plant V_1 then p_1L is the expected loss from an infestation originating on the farmer's own land and $(1 - p_1)\, p_1qL$ is the expected loss from an infestation arising on the other farmer's land (probability p_1), transferring across farms (probability q), all conditioned on there not having been an infestation already originating from the home plot $(1 - p_1)$.[3] We can represent these payoffs in the conventional matrix, where the rows are F_1's strategy and the columns F_2's strategy:

	V_1	V_2
V_1	$Y - p_1L - (1 - p_1)p_1qL, \ldots$	$Y - p_1L,\ Y - p_2L$
V_2	$Y - p_2L,\ Y - p_1L$	$Y - p_2L - (1 - p_2)p_2qL, \ldots$

Recall that all p's and q are probabilities so that they are between zero and one. From this matrix it is easy to show that there will be a Nash equilibrium at which both choose different crops if $p_1 < p_2(1 + q) - q(p_2)^2$ and $p_2 < p_1(1 + q) - q(p_1)^2$. For $q = 0$, these two inequalities are inconsistent. For $q > 0$, they are consistent for $p_1 = p_2$ and by continuity for an open set of p values containing the diagonal. They are consistent for:

$$q > \max\left\{\frac{p_1 - p_2}{p_2\left(1 - p_2\right)}, \frac{p_2 - p_1}{p_1\left(1 - p_1\right)}\right\}$$

So, in general, there is a set of values of p_1 and p_2 having positive measure that are consistent with a Nash equilibrium at which the two farmers choose different crops. Fig. 28.2 illustrates this: the region in which diversity is an equilibrium is lens-shaped and increases as q increases.

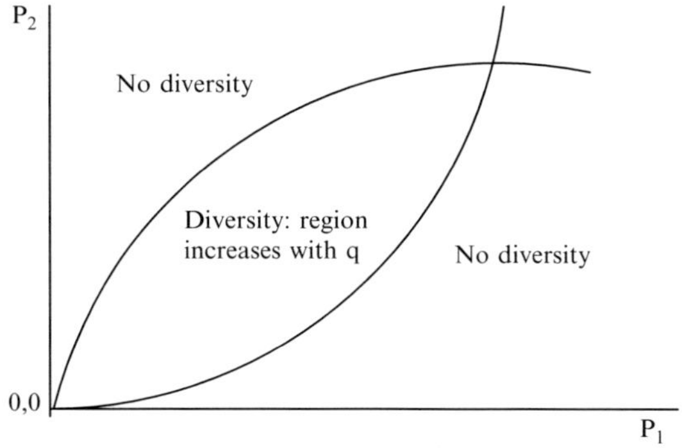

Fig. 28.2 How diversity at equilibrium depends on the parameters

[3] This is an illustration of the interdependent security problem discussed by Kunreuther and Heal (2002) and Heal and Kunreuther (2002).

It is also possible to show that in the area outside the lens, both farms will use the same crop at a Nash equilibrium. This area satisfies the reverse inequalities to those given in paragraph above, namely $p_1 > p_2(1 + q) - q(p_2)^2$ and $p_2 > p_1(1 + q) - q(p_1)^2$. We can also tie this analysis back in to the risk–yield frontier considered earlier (Fig. 28.1). If both farmers choose different varieties then their expected incomes are $Y - p_1L$ and $Y - p_2L$ for varieties 1 and 2, respectively, and the variances of their returns are $p_1(1 - p_1)L^2$ and $p_2(1 - p_2)L^2$. If, however, both choose the same variety then the expected incomes are $Y - p_1L - (1 - p_1)p_1qL$ or $Y - p_2L - (1 - p_2)p_2qL$ depending on the common crop and the variances are $[1 - p_1 - (1 - p_1)p_1q][p_1 + (1 - p_1)p_1q]L^2$ or the same with 1 replaced by 2. It is now clear that for each crop the expected return is lower and the variance of the return higher when both farmers plant it than when only one plants it. So, the risk–return frontier moves down and to the right as we move from considering an isolated farmer to a farmer who is surrounded by others choosing the same crop.

28.4 Policy Responses

The existence of external effects between farmers implies that the overall allocation of risks in society, and indeed the overall allocation of resources, will be inefficient. The analysis of Nash equilibria above indicates that under a range of parameter values farmers will make identical choices of variety and will select too little diversity. To achieve efficiency we would need to face each farmer with a cost that indicates the extent of the external effect that he is imposing on others by making the same choice as them. The result would be that individuals would to some degree be steered away from making the same choices as their neighbors, and so the resulting level of genetic diversity in food crops and animals would be greater than at present. In terms of the diagram, we wish to expand the lens area to include all likely values of p and q.

There are several institutional mechanisms through which this could be effected. The classical responses to external effects are either to levy taxes or to redefine property rights. In the present case, the tax-based approach would require a tax that depends on choice of variety – a homogeneity tax. In practical terms, we would look for a tax rate on seeds that is higher on the more popular varieties and so provides an incentive to move to the less popular ones. Equivalently, we could think of a subsidy to the less popular varieties, bringing down their prices and again transferring demand to them (Ehrlich and Ehrlich 1982). To the extent that the obstacles to the use of the less popular varieties are lack of knowledge and lack of access to advice and support, rather than purely financial issues, then the appropriate response would also involve increased provision of information about the use of the less popular varieties. A solution based on reform of property rights would be more complex: typically Coasian approaches to externalities have given rise to "cap and trade" mechanisms such as that used for SO_2 in the US and proposed by the Kyoto Protocol for the control of greenhouse gases. For varietal choice, such a mechanism might work by requiring permits for the use of the more popular varieties and then limiting the numbers of permits issued and making them tradable. There is currently a requirement imposed on users of *Bt* corn by its producer, Monsanto, that is a crude approximation to these measures: farmers who plant *Bt* corn are required to plant certain fraction of their land with non-*Bt* products as a way of limiting the development of pathogens that are resistant to the toxins expressed by the *Bt* genes. This way they hope to maintain the effectiveness of the *Bt* toxin as an insecticide. The use of *Bt* corn raises several issues, including the possible impacts of plant-produced toxins on wildlife and the possible loss of effectiveness of the *Bt* toxin, and this latter issue is close to those considered in this paper. Concentration of large areas under *Bt* corn will in the long run reduce its effectiveness and will also reduce the effectiveness of some of the benign insecticides traditionally used by organic farmers. There is, therefore, a social interest in crop variety, though individual farmers may find that from a private perspective their best choice is to use only the *Bt* varieties. That this problem has been recognized and is being tackled in this limited area is encouraging, and sets a precedent for the types of measures required on a larger scale.

28.5 Conclusions

In summary, genetic diversity in agriculture is important. Its lack can increase social vulnerability to pathogens and increase the risks that we face with our food supplies. Genetic diversity in agricultural systems is changing, and there has been a decline in traditional crop varieties and in breeds and varieties of animals. This may be associated with growing international trade in foods and in seeds. There is now perhaps an increase in the diversity of new genes included in new varieties, but the net effect of this is uncertain. A consequence of these changes is a potentially serious increased risk of food production failure. The solution lies in recognizing the risk, specifying regional goals to contain it, and facilitating appropriate institutions (or international rules) to achieve them.

Whatever the historical trends in agricultural genetic diversity, and there is some argument about these, there are strong a priori reasons to think that individual farmers will choose a degree of genetic homogeneity that is greater than is socially optimal, and that there is consequently a case for policy intervention here. This observation would not be invalidated by a flat or even increasing trend in agricultural biodiversity: in the case of an increasing trend it would imply that for a socially efficient outcome the upward trend should be even stronger. We have indicated the types of interventions that might be appropriate. The problem that we have identified has features in common with one of the problems arising from the use of *Bt* corn and the desire to maintain the effectiveness of the *Bt* toxin as an insecticide. A problem with a similar structure arises with respect to the development of antibiotic resistance in bacteria. Here, the private and social costs of the choice of an antibiotic are different, and in particular each time an antibiotic is used this imposes costs on other who might need to use it.

References

Bellon, M. R. (1996). The dynamics of crop intraspecific diversity: A conceptual framework at the farmer level. *Economic Botany, 50*, 26–39.

Browning, J. A. (1988). Current thinking on the use of diversity to buffer small grains against highly epidemic and variable foliar pathogens: Problems and future prospects. In *Breeding strategies for resistance to the rusts of wheat* (pp. 76–90). Mexico, DF: CIMMYT.

Bush, S. B. (Ed.). (2000). *Genes in the field: On-farm conservation of crop diversity*. London: Lewis.

Committee on Agriculture Science and Technology. (2003). Benefits of Biodiversity.

Ehrlich, P., & Ehrlich, A. (1982). Ecoscience: Preserving genetic diversity. Seed banks. *Mother Earth News*, 146–147.

Hammer, K., Knupffer, H., Xhuveli, L., & Perrino, P. (1996). Estimating genetic erosion in landraces – Two case studies. *Genetic Resources and Crop Production, 43*, 329–336.

Heal, G., & Kunreuther, H. (2002). *You only die once: Managing discrete interdependent risks*. Columbia Business School.

Kunreuther, H., & Heal, G. (2002). Interdependent security. *Journal of Risk and Uncertainty, 26*(2/3), 231–249. Special issue on Terrorist Risks.

Lopez, P. B. (1994a). A new plant disease: Uniformity. *CERES, 26*(6), 41–47.

Marshall, D. R. (1977). The advantages and hazards of genetic homogeneity. *Annals of the New York Academy of Sciences, 287*, 1–20.

Myers, N. (1997). Biodiversity's genetic library. In G. Daily (Ed.), *Nature's services: Societal dependence on natural ecosystems*. Washington, DC: Island Press.

Porceddu, E., Ceoloni, C., Lafiandra, D., Tanzarella, O. A., & Scarascia Mugnozza, G. T. (1988). Genetic resources and plant breeding: Problems and prospects. In T. E. Miller & R. M. D. Koebner (Eds.), *Proceedings of the seventh international wheat genetics symposium* (pp. 7–22). Cambridge: Institute of Plant Science Research.

Smale, M. (1997). The green revolution and wheat genetic diversity: Some unfounded assumptions. *World Development, 25*(8), 1257–1269.

Smale, M. (1998a). The contribution of genetic resources and diversity to wheat production in the Punjab of Pakistan. *American Journal of Agricultural Economics, 80*(3), 482–493.

Smale, M. (Ed.). (1998b). *Farmers gene banks and crop breeding: Economic analysis of diversity in wheat maize and rice*. Boston: Kluwer Academic.

US National Research Council. (1993). Managing global genetic resources: Agricultural crop issues and policies.

Wood, D., & Lenne, J. M. (1997). The conservation of agrobiodiversity on-farm: Questioning the emerging paradigm. *Biodiversity Conservation, 6*, 106–120.

Chapter 29
Askö 2000: Commentary by Alison G. Power

Alison G. Power

The conservation of crop genetic diversity has been of significant concern to plant breeders and the agricultural community for decades. There is broad consensus that conserving genetic diversity is an important societal goal. Crop genetic diversity is advantageous in the short-term, because it can limit epidemics of crop diseases (Mundt 2002; Finckh 2008). Crop genetic diversity is advantageous in the long-term, because it provides a resource essential to plant breeding. Debate has focused on the extent and causes of genetic erosion, as well as the advantages and disadvantages of different methods of crop conservation, including *in situ* conservation by farmers and *ex situ* conservation in gene banks. The 2008 opening of the Svalbard Global Seed Vault on the Norwegian island of Spitsbergen, intended as the ultimate safety net against accidental loss of crop genetic diversity, recently brought the conservation issue to international attention.

The deployment of genetic diversity by farmers has important implications for genetic conservation. While some farmers, particularly those with few resources, may choose agricultural diversity as a risk–averse strategy in the face of economic and environmental uncertainty, many farmers select crops and cultivars to maximize yields and market returns. As Heal et al. highlight, the choices that farmers make in order to balance yields and acceptable risk may not result in maximizing the public good of genetic conservation. Moreover, farmers' choices may not even maximize their own returns, since their neighbors' decisions will influence the risk they incur. Under many scenarios farmers are likely to choose the same crop variety, and the result of many farmers' cultivar selections will be too little genetic diversity (Heal et al.).

There are a variety of mechanisms that could be employed to increase crop genetic diversity in farmers' fields and contribute to genetic conservation. Heal et al. offer a number of possible policy alternatives to increase the genetic diversity utilized by farmers, including taxes on popular varieties, subsidies for unpopular varieties, and cap and trade mechanisms. Institutions such as local marketing and

---

A.G. Power
Department of Ecology and Evolutionary Biology, Cornell University, USA
e-mail: agp4@cornell.edu

T. Söderqvist et al. (eds.), *Bringing Ecologists and Economists Together*,
DOI 10.1007/978-90-481-9476-6_29,

processing cooperatives may also contribute to the conservation of genetic diversity (Di Falco et al. 2008). In addition to these social and economic mechanisms, there are also a number of biological approaches to retaining crop genetic diversity, some of which have already been widely adopted in major crops.

Genetic diversification can occur at several levels within and among cultivars. Within cultivars, useful genetic diversity can be retained using breeding approaches that include composite crosses, random-mated populations, mixtures of two- and three-way crosses, and dynamic gene pool management with farmer-participatory selection. Genetic mixtures in the form of multiline cultivars have been used in small grains such as oats since the 1960s (Browning and Frey 1969; Wolfe 1985). These cultivars are composed of lines bred for agronomic uniformity that differ only in disease resistance traits. In the 1980s cultivar mixtures, composed of agronomically compatible cultivars that differ in resistance genes, became increasingly popular. Cultivar mixtures require no additional breeding and can be modified readily as new cultivars become available (Mundt 2002). Since cultivar mixtures have greater genetic diversity than multilines, they may provide protection from a greater range of diseases and pests.

Both multiline cultivars and cultivar mixtures are commercially produced and sold by large seed companies. Multilines have been used for decades to control rust in U.S. oat production. Barley cultivar mixtures have been widely used to control powdery mildew in western and central Europe and now make up approximately 60% of the area sown with spring barley (Finckh 2008). Wheat cultivar mixtures are grown regularly in the Northwestern U.S., comprising 13% and 18% of current wheat area of Oregon and Washington, respectively (Mundt 2002). Cultivar mixtures have been successfully used over a large area in China to control rice blast in glutinous rice (Zhu et al. 2000).

It is clear that biological, institutional, and policy approaches can all contribute to genetic diversification that can benefit farmers through disease and pest control, while at the same time conserving crop genetic diversity for future generations. Understanding the factors that influence farmers' choices of cultivars and cropping patterns is essential to shaping the right combination of approaches for a particular crop and socioeconomic context. The analysis of Heal et al. will aid in designing genetic conservation strategies for important crops.

References

Browning, J. A., & Frey, K. J. (1969). Multiline cultivars as a means of disease control. *Annual Review of Phytopathology, 7*, 355–382.

Di Falco, S., Smale, M., & Perrings, C. (2008). The role of agricultural cooperatives in sustaining the wheat diversity and productivity: The case of southern Italy. *Environmental and Resource Economics, 39*, 161–174.

Finckh, M. R. (2008). Integration of breeding and technology into diversification strategies for disease control in modern agriculture. *European Journal of Plant Pathology, 121*, 399–409.

Mundt, C. C. (2002). Use of multiline cultivars and cultivar mixtures for disease management. *Annual Review of Phytopathology, 40*, 381–410.

Wolfe, M. S. (1985). The current status and prospects of multiline cultivars and variety mixtures for disease resistance. *Annual Review of Phytopathology, 23*, 251–273.

Zhu, Y. Y., Chen, H. R., Fan, J. H., Wang, Y. Y., Li, Y., Chen, J. B., et al. (2000). Genetic diversity and disease control in rice. *Nature, 406*, 718–722.

Chapter 30
Askö 2000: Commentary by Stephen Polasky

Stephen Polasky

Conserving biodiversity is not just about saving tropical rainforests, coral reefs, and charismatic megafauna. Conserving biodiversity is also about maintaining genetic diversity within species. From a human welfare perspective, maintaining genetic diversity of major agricultural crops has large potential value. One reason to conserve such genetic diversity is to provide insurance against bad outcomes. A monoculture crop may be highly productive but may also be highly susceptible to disease or ill-suited to some range of possible weather conditions (drought, temperature extremes). A large closely linked host population tends to increase the likelihood of the evolution of a pathogen capable of exploiting the host, and facilitates the spread of the pathogen throughout the population. Planting high yield but drought intolerant varieties may generate a high yield on average but provides little protection against disaster should drought occur. When facing risks, diversifying your portfolio is standard financial advice, often expressed as a biological metaphor of "not putting all your eggs in one basket." Similarly, planting multiple varieties of a crop, or multiple crops, can reduce the probability of large losses from disease, either because of slower evolution of pathogens or because of decreased ability to spread, or large losses from extreme weather conditions. In fact, when conditions are highly variable across space or time, a diverse cropping pattern can be both more productive as well as more stable because it allows for more complete utilization of resources across heterogeneous conditions (Tilman et al. 2005).

The spread of modern agricultural techniques has tended to reduce crop diversity and varietal diversity within major crops. Given the potential benefits of genetic diversity, an important question is whether adequate genetic diversity remains or whether too much diversity is being lost. Using a simple model that accounts for potential spread of disease, Heal et al. show that farmers planting the same crops impose a negative externality on each other. Without some type of policy intervention or negotiation among farmers, farmers will have too much incentive to plant the same crops and too little incentive to plant a diverse set of crops. Brock and

S. Polasky
Department of Ecology, Evolution, and Behavior, University of Minnesota, USA
e-mail: polasky@umn.edu

T. Söderqvist et al. (eds.), *Bringing Ecologists and Economists Together*,
DOI 10.1007/978-90-481-9476-6_30, © Springer Science+Business Media B.V. 2011

Xepapadeas (2003) demonstrate a similar point in an elegant dynamic model that incorporates the evolution of pathogens.

There are good reasons for thinking that individual farmers do not have an adequate incentive to maintain genetic stock in the face of potential evolution or spread of pests and pathogens. But not all threats to crop yields necessarily generate inadequate incentives to maintain diversity. Suppose the risk facing farmers is the threat of drought rather than disease. Farmers can plant either a standard variety that has high yields under normal conditions, but low yield in drought, or a drought resistant variety that has moderate yields under all conditions. In drought conditions when supplies are limited prices for crops are likely to be high. It may be privately, as well as socially, optimal for farmers to grow some drought resistant varieties to take advantage of the potential high prices if drought occurs and some standard varieties to take advantage of large yields under normal weather conditions. In such cases, there need not be an excessive tendency toward monoculture.

Genetic diversity in agricultural crop species is an important capital asset. Heal et al. provide a valuable service by highlighting the issue of whether there are sufficient incentives to maintaining this genetic diversity. The paper shows reasons for believing that private incentives are insufficient for adequate maintenance of genetic diversity in agriculture. However, more work is necessary to provide answers on the best type of policy response as well as quantitative levels of such a response. In principle, the optimal policy response might be quite complex, such as a spatially differentiated tax that reflects the negative externalities to neighboring farmers planting the same crop. Such taxes would be difficult to institute and administer. Simpler tax or regulatory schemes that are not spatially differentiated would not generate an optimal solution but might be more politically acceptable and easier to implement. An alternative type of policy, rather than attempting to change incentives facing farmers, is to have a publicly supported seed banks to maintain genetic diversity. The Consultative Group on International Agricultural Research (CGIAR) maintains a collection with over 600,000 samples. Ex situ storage does not address the externality from spread of disease but may assist with finding varieties capable of resisting existing or emerging diseases. Besides thinking about the type of policy intervention, there is a need for empirical research that integrates natural science understanding of yields, disease threats, genetics and evolution, with economic costs and benefits to demonstrate the degree of risk and the benefits of maintaining genetic diversity.

References

Brock, W. A., & Xepapadeas, A. (2003). Valuing biodiversity from an economic perspective: A unified economic, ecological and genetic approach. *American Economic Review, 93*(5), 1597–1614.

Tilman, D., Polasky, S., & Lehman, C. (2005). Diversity, productivity and temporal stability in the economies of humans and nature. *Journal of Environmental Economics and Management, 49*(3), 405–426.

Chapter 31
Askö 2001: Sustainability's Compass – Indicators of Genuine Wealth*

Anastasios Xepapadeas, Steve Carpenter, Sara Aniyar, Kenneth Arrow, Gretchen Daily, Partha Dasgupta, Paul Ehrlich, Carl Folke, Geoff Heal, Michael Hoel, Nils Kautsky, Simon Levin, Jane Lubchenco, Karl-Göran Mäler, Elinor Ostrom, Thomas Rosswall, David Starrett, and Brian Walker

*This article was published in *Beijer Discussion Paper Series No. 188*, Xepapadeas et al., Sustainability's compass: indicators of genuine wealth, Copyright the Beijer Institute of Ecological Economics (2004).

A. Xepapadeas
Department of Economics, University of Crete, Rethymno 74100, Crete, Greece

S. Carpenter
Center for Limnology, University of Wisconsin, Madison, WI 53706, USA

S. Aniyar and K-G. Mäler
Beijer Institute of Ecological Economics, the Royal Swedish Academy of Sciences, Box 50005, SE-10405 Stockholm, Sweden

K. Arrow and D. Starrett
Department of Economics, Stanford University, Stanford, CA 94305, USA

G. Daily and P. Ehrlich
Department of Biological Sciences, Stanford University, Stanford, CA 94305, USA

P. Dasgupta
Faculty of Economics, Cambridge University, Sidgwick Avenue, Cambridge CB3 9DD, UK

C. Folke and N. Kautsky
Department of Systems Ecology, Stockholm University, S-10691 Stockholm, Sweden

G. Heal
Program on Information and Resources, Columbia University, New York, NY 10027, USA

M. Hoel
Department of Economics, University of Oslo, Box 1095, Blindern, 0317 Oslo, Norway

S. Levin
Department of Ecology and Evolutionary Biology, Princeton University, Princeton, NJ 08544, USA

J. Lubchenco
Department of Zoology, Oregon State University, Corvallis, OR 97331, USA

T. Söderqvist et al. (eds.), *Bringing Ecologists and Economists Together*,
DOI 10.1007/978-90-481-9476-6_31, © Springer Science+Business Media B.V. 2011

E. Ostrom
Workshop on Political Theory and Policy Analysis, Indiana University, Bloomington, IN 47403, USA

T. Rosswall
International Foundation for Science, Grev Turegatan, 19, 114 38 Stockholm, Sweden

B. Walker
CSIRO Sustainable Ecosystems Division, GPO Box 284, Canberra, ACT 2601, Australia

31.1 Introduction

Sustainability of the Earth's social, economic and ecological systems concerns the status, trends and future prospects for these systems as a whole (National Academy of Sciences 2000). Preserving or enhancing the flows of goods and services that humans derive from ecosystems is a condition for sustainability. These flows are multidimensional and complex, with many tradeoffs that are poorly understood. Yet, humanity must make decisions now that will affect ecosystem goods and services for decades into the future. These decisions will rely on indicators of future condition of the whole social, economic and ecological system, but how should these indicators be chosen? Economics and ecology converge on indicators that represent forward-looking measures of the capacity of ecosystems to supply a non-decreasing flow of goods and services.

31.2 Economics: Non-declining Wealth

Non-declining genuine wealth is a consequence of a simple, yet illuminating, definition of sustainability (Dasgupta and Mäler 2000). Genuine wealth, defined in an inclusive way, is the accounting worth of an economy's capital assets, with the capital base consisting of produced, human, and natural capital (Arrow et al. 2003; Dasgupta and Mäler 2000, 2001).

Importantly, this concept includes the future capacity of ecosystems to support human well-being. Changes in genuine or inclusive wealth, measured in constant accounting prices, represent genuine investment, or net change in capital (Dasgupta and Mäler 2001; World Bank 1997). This is because accounting prices reflect the discounted value of the future flow of changes in welfare due to changes in the stock of the capital base (Dasgupta and Mäler 2001). A basic advantage of genuine wealth over other commonly used indices such as GDP or the Human Development Index, is that wealth (by using accounting prices to value changes in the capital base) is a forward-looking measure that reflects expected future changes in well-being. For example, wealth is related to measures of literacy and education that reflect the future productive capacity of people (Dasgupta and Mäler 2001).

The capital base, however, is highly heterogeneous, including combinations of tangible and intangible items, some of which have no markets (Heal 1998; National Academy of Sciences 1999). This presents a major difficulty in measuring genuine wealth, although, as shown below, it is not impossible to overcome these difficulties in order to produce some approximations. With the exception of produced capital, the market does not supply forward-looking measures of social value that reflect accounting prices. One of these problematic components of genuine wealth is natural capital, such as that associated with biodiversity or the capacity of ecosystems to supply goods and services despite environmental shocks.

The notion of wealth distinguishes between the current productive capacity of a system and its capacity to produce in the future. The future productive capacity of the system is closely associated with the concept of resilience (Levin et al. 1998; Gunderson 2000; Gunderson and Holling 2001). More resilient systems are less likely to be degraded to configurations that reduce inclusive wealth.

31.3 Ecology: Non-decreasing Resilience

Resilience is measured by the change that can be tolerated before an ecosystem moves to a different configuration, controlled by different processes and yielding a different flow of ecosystem services (Carpenter et al. 2001; Folke et al. 2002). Once such a shift has occurred, restoration of the original condition may be expensive or impossible. Massive shifts have been documented for freshwater, marine, rangeland, forest and arctic ecosystems (Folke et al. 2002; Scheffer et al. 2001). Frequently such shifts in ecosystems have severe impacts on natural capital and flows of ecosystem goods and services.

While each shift in ecosystem services has unique site-specific characteristics, there are general patterns that arise in case after case (Scheffer et al. 2001). Among these are: slowly-changing variables that control the basins of attraction for various ecosystem states; feedbacks that can rapidly change direction at critical thresholds; and stochastic shocks that can trigger rapid collapse when resilience is sufficiently low. Often the slowly-changing variables provide clues to resilience.

For example, resilience of lake water quality is related to concentrations of phosphorus in soils and sediments, which change more slowly than phosphorus concentrations in lake water and provide a useful leading indicator of water quality (Carpenter et al. 2001). Lake eutrophication is a familiar example (Box 31.1). For the ecological-economic system composed of a lake and its agricultural catchment, one can study the optimal control of ecosystem service flows – inclusive wealth as defined above. For a wide range of realistic conditions, the optimal trajectories possess different characteristics. There could be monotonic convergence or cyclic convergence to an oligotrophic or a eutrophic steady state depending on initial conditions (Mäler et al. 2003; Carpenter et al. 1999a, 1999b). However, the clear water state can be maintained for long periods of time by management

Box 31.1 Lake Eutrophication

Eutrophication is a state of lakes characterized by toxic blooms of algae, episodes of anoxia, fish kills, and loss of freshwater ecosystem services such as drinking water, water for irrigation and industrial use, fisheries, and recreation (Postel and Carpenter 1997). The fundamental cause is enrichment with phosphorus (P) (Reed and Carpenter 2002). P is an element in agricultural fertilizers and livestock feed which tends to accumulate in soil (Fig. 31.1a). Erosion washes the soil into streams and lakes, where it dissolves and becomes available to algae, including toxic bloom-forming species which are favored at high P concentrations. P also accumulates in sediments. It is rapidly recycled from sediments during episodes of anoxia, which are caused by decay of dead algae.

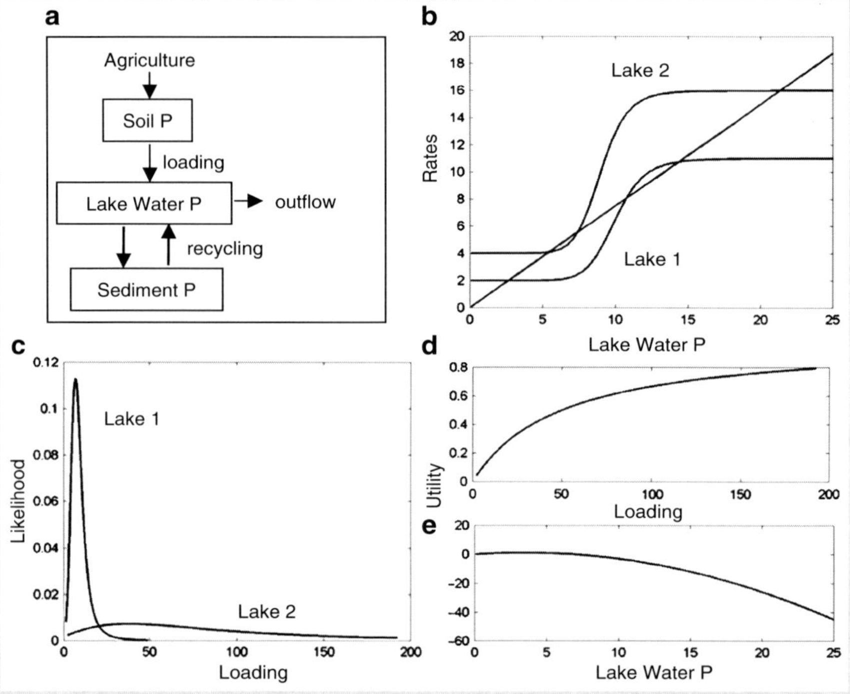

Fig. 31.1 (**a**) Flows of phosphorus (P) from agricultural sources (fertilizer, manure) into soil, lake water, and sediments. (**b**) Rates of P flow into and out of lake water, as a function of P mass in the water. Straight line is loss flux by hydrologic outflow and sedimentation. Curved line is mean loading from the watershed plus recycling from sediments and consumers. Lake 1 has low mean loading and low recycling, lake 2 has high mean loading and high recycling. (**c**) Probability distributions of annual load for Lake 1, with low mean loading, and Lake 2, with high mean loading. (**d**) Utility generated annually by agricultural activities that cause loading of P to the lake. (**e**) Utility generated annually by ecosystem services such as fresh water for human use, fish, and recreation

(continued)

Box 31.1 (continued)

A minimal model of P dynamics in lakes (Fig. 31.1b) includes processing of P by outflow and sedimentation (straight line), and inputs plus recycling (curved lines). Two hypothetical lakes are shown in Fig. 31.1b. Lake 2 has higher average input from the soil (intercepts of curved lines) and a higher maximum rate of recycling (height of curved lines). Both lakes have identical outflow and sedimentation. Steady states occur where the straight and curved lines intersect. Both lakes have three steady states. The middle one is an unstable repeller, and the upper and lower ones are locally stable attractors. The steady state at low P concentration is the clear-water state, and the steady-state at high P concentration is the turbid or eutrophic state.

Resilience of a state is measured by the distance (in P units) from a stable steady state to the unstable steady state. The resilience of the clear-water state is greater for lake 1 than for lake 2. Resilience is related to the size of a stochastic shock that the system can absorb without changing states. A shock which moves the system from one stable state past the unstable steady state will shift the system to the other stable state. For example, a P input event that is larger than the resilience of the clear-water state will push the lake past the unstable steady state, and recycling will then drive the P level to the eutrophic steady state.

Soil P affects the probability distribution of input events to the lake (Fig. 31.1c). These input events are driven by variable and unpredictable storms and snowmelt, so we model them as a random variable. The logarithm of inputs usually fits a Student-t or gamma distribution. The watershed of lake 2 has high-P soils causing a higher mean input rate (Fig. 31.1b) and broader distribution of stochastic shocks (Fig. 31.1c). Therefore, there is a relatively high probability of a shock large enough to move lake 2 out of the clear-water state.

Utility is derived from activities that cause P loading to the lake, such as intensive use of fertilizers or animal-intensive farming. These utilities increase (with decreasing slope) as P loading rate increases (Fig. 31.1d). Utilities are also associated with water quality as measured by the amount of P in the lake water. These utilities are ecosystem services that may not have markets, such as water for diverse human uses, fisheries, or recreation. These utilities increase with P and low P levels (Fig. 31.1e), because a small amount of P supports production of ecosystem services such as fishes and waterfowl. Above an optimal level of P in the lake, however, utility declines.

according to practical, measurable indicators of resilience. Soil P concentration should be kept relatively low, and the target level for soil P is lower for lakes that have more P in the sediments (hence greater potential recycling). A simple way to stabilize soil P is to create incentives for farm P budgets that balance (i.e. imports of P in fertilizer and animal feed equal exports of P in grain, produce, dairy products, eggs and meat). For example, farm exports of P (in the form of food) could be tied to marketable credits for purchase of P (such as fertilizer or animal feed) (Scheffer et al. 2001).

In the lake example, the accounting price for P reflects the present value of a change in well-being associated with the specific lake system resulting from a change in the P concentration. Thus if accumulated P (or its inverse) is regarded as our environmental capital, its value can be measured locally using a fixed accounting price. If we take a global approach, then accounting prices change, and their change should provide the signal for the direction of change of inclusive wealth for the lake system. Thus in the lake case, although the system is fairly complex, our understanding of the physical and economic processes involved allows us to approximate the wealth concept as a proxy for well-being associated with the lake system. In this system we obtain two interrelated types of signals regarding the possible loss of resilience of the system because of perturbations, which could be reflected in a fast movement from a desirable oligotrophic state to an undesirable eutrophic state. One is the concentration of P: an increase in P could indicate that the system is moving towards a threshold point after which it would move fast to a eutrophic basin of attraction. At the same time the accounting price for P should change, indicating that the move to the eutrophic region will reduce the present value of the benefit flow from the lake (Ludwig et al. 2003). Thus changes in the P concentration or its accounting price can be regarded as sufficient indicators of the resilience of the flow of services from the lake system, to perturbations. We want these indicators to be forward looking. In the lake example, indicators based on soil P will be more successful in having the forward looking property than those based on water. On the other hand, accounting prices are forward looking by construction, although admittedly much harder to estimate.

Similarly, the resilience of rangelands is related to dynamics of woody vegetation, which changes more slowly than grass biomass and serves as a leading indicator of rangeland productivity (Carpenter et al. 2001). In these well-studied cases, we know enough about the processes driving the system to devise indicators. Because this is not always the case, data collection as part of the indicator-building task could be used in learning the structure of ecosystem processes and devising better indicators of wealth associated with natural capital.

In general, the ecological variables that control resilience should be kept within certain bounds, and these bounds change depending on other variables that control critical feedbacks. The probability of a state shift also depends on the disturbance regime, which is affected by ecosystem properties as well as events outside the ecosystem. Despite the enormous differences in detail among ecosystems, the general challenge is the same: to avert a slow loss of resilience followed by a stochastic shock that leads to a long-lasting change in ecosystem service flows.

Indicators that provide the signals needed to maintain resilience and reduce the risk of long-lasting losses of ecosystem goods and services are surrogates for the natural capital component of wealth. Wealth indicators must be tied to slowly-changing ecological variables that govern resilience. In this sense, ecological wealth can be thought of as providing insurance against possible shocks that move the system to an undesirable state and, as a result, might reduce the flow of useful ecosystem services.

In the lake case a clear lake – low P concentrate – can be thought of as indicating a high value of ecological wealth. This wealth – clear state – can be thought of as providing insurance against a possible shock that causes an increase in the P concentrate, in the sense that it prevents the system from going over the threshold, losing its resilience, and moving towards an undesirable eutrophic basin of attraction. Similar examples of a relationship between wealth in the form of environmental capital, providing insurance against loss of valuable ecosystem services due to perturbations, and resilience can be found in cases including forests, fish biomass, wetlands, and so forth. Correspondingly, the accounting prices for ecological resilience can be thought of as the insurance premium we are willing to pay for this insurance.

31.4 Forward-Looking Indicators

Wealth, accounting prices, and ecological resilience are all forward-looking by design. They reveal the guardrails within which complex social and ecological systems can sustain a particular level of human well-being. While the theoretical meaning of inclusive wealth is clear, precise calculation of wealth is difficult, especially when we want to capture the forward-looking aspects that we seek. A pragmatic approach is needed, analogous to the various indicators used by governments to gauge economic activity, or by securities markets to evaluate performance of investments. Society routinely bases important decisions upon, and pays considerable sums to measure, indicators such as GDP or the various stock-market indices. Such indices are known to be insufficient statistics but nevertheless provide a well-understood basis for comparison across nations and across years. Natural capital is surely worthy of comparable attention. An appropriate set of wealth indicators could help guide humanity toward sustainability.

Recent estimates of genuine wealth (Arrow et al. forthcoming), where changes in natural capital are reflected in CO_2 damages, energy minerals and forest depletion, show a striking difference between the picture that emerges of the development path of an economy when genuine wealth per capita is used as a yardstick instead of the conventional GDP per capita. As shown in Table 31.1, column 4, there are economies which cannot be regarded as sustainable because genuine wealth per capita, before adjustments for changes in total factor productivity (TFP) is taken into account, is declining, while for the same economies GDP per capita is increasing. When technological change is accounted for through TFP adjustments, there is an overall improvement regarding the sustainability of some individual economies as shown in column 3 of Table 31.1. It should be noted however that the contribution of TFP to the sustainability criterion of genuine wealth growth is expected to be overestimated. This is because conventional TFP estimates are likely to be biased upwards because they do

Table 31.1 Growth rate of per-capita genuine wealth and per capita GDP

Country	Genuine investment as percent of GDP (1)	Growth rate of per-capita genuine wealth – before TFP adjustment (2)	Growth rate of per-capita genuine wealth – after TFP adjustment (3)	Growth rate of per capita GDP (4)
Bangladesh	7.09	−1.09	0.30	1.88
India	9.46	−0.57	0.54	2.96
Nepal	13.31	−0.24	0.63	1.86
Pakistan	8.75	−1.35	0.59	2.21
China	22.73	2.06	8.33	7.77
Sub-Saharan Africa	−2.08	−3.05	−2.58	−0.01
Middle East and North Africa	−7.09	−3.43	−3.82	0.74
United Kingdom	7.38	1.30	2.29	2.19
United States	8.94	0.72	0.75	1.99

Source: Arrow et al. (forthcoming)

not take into account the depletion of environmental capital (Dasgupta and Mäler 2000, Xepapadeas forthcoming).

In estimating particular indices, transparency is essential; the procedure for calculating the index must be well-understood and repeatable by all. The index will necessarily be measured by sampling an arbitrary (but clearly stated and widely-agreed-upon) set of ecosystems, of specified spatial extent and distribution, at specified time intervals.

Indices that represent future changes in natural capital will be instrumental in enhancing the insurability of environmental risks. Liability insurers are highly exposed to environmental risks, which are characterized by ambiguous probability that an environmental loss will occur and by ambiguity of the value of the loss itself (Swiss Re 1998). By using early warning indicators to characterize the adaptive capacity of ecosystems, the probability of losses will become less ambiguous. Furthermore, by valuing environmental assets using forward-looking accounting prices, the ambiguity about the value of the loss will also be reduced. The reduction in ambiguity leads to a reduction in premiums. The effect of wealth indicators on insurance premiums is analogous to those of the Eco-Management and Audit Scheme (EMAS) or ISO 14000. Eventually, these reductions in ambiguity will reduce the vulnerability of the re-insurance sector (defined as the relationship between probable maximum losses, the probability of the catastrophic event, and the sector's capacity to pay for these losses).

Substitutability is a critical issue in designing an index of inclusive wealth. For market goods, pricing reflects the tradeoffs that people are willing to make between different priorities, for example food and shelter. How can such tradeoffs be assessed for fundamentally different aspects of ecological wealth, such as the future supply of soil fertility and water quality, or the future supply of a predator and its prey? At what spatial scale should ecological wealth be assessed – nations, major river basins, the world? How should wealth be aggregated across spatial units or different types of ecosystems? This is an area of great importance for future research. It is clear that in dealing with ecological wealth, one has to accept conditions of limited substitutability, or thresholds above which substitution is not possible. Until these tradeoffs are understood, a variety of disaggregated indicators should be employed that represent diverse aspects of ecosystem wealth.

Ecosystem resilience implies some pragmatic solutions to these problems. For example, freshwater ecosystem resilience and agricultural production could be maintained by balancing farm phosphorus budgets through market mechanisms, without detailed understanding of site specific complexities of the phosphorus cycle or the individual decisions of millions of farmers (Bennett et al. 2001).

Similar rules of thumb can be devised for other types of ecosystems, where the foundations of resilience may lie in wilderness and unconverted ("undeveloped") land and habitat; diversity of crops and animals in the food base (including genetic diversity), species biodiversity, and diversity of habitats; institutions that facilitate conservation by individual landowners, local people, and government at multiple levels; and markets to capture externalities such as emissions of pollutants, fertilizers, pesticides and herbicides.

Activities to build ecosystem wealth can be initiated now. If action is combined with measurements of indicators of inclusive wealth, society will establish the empirical baseline needed to improve management of our common wealth in the future.

References

Arrow, K. et al. (forthcoming). Are we consuming too much? *Journal of Economic Perspectives*.

Arrow, K., Dasgupta, P., Mäler, K.-G. (1993). Evaluating projects and assessing sustainable development in imperfect economies. *Environmental and Resource Economics, 26*, 647–685.

Bennett, E. M., Carpenter, S. R., & Caraco, N. F. (2001). Human impact on erodable phosphorus and eutrophication: A global perspective. *BioScience, 51*, 227.

Carpenter, S. R., Brock, W. A., & Hanson P. C. (1999a). Ecological and social dynamics in simple models of ecosystem management. *Conservation Ecology 3*(2):4, [online] URL http://www.conscol.org/vol3/iss2/art4.

Carpenter, S. R., Ludwig, D., & Brock, W. A. (1999b). Management of eutrophication for lakes subject to potential irreversible change. *Ecological Applications 9*, 751–771.

Carpenter, S. R., Walker, B. H., Anderies, J. M., & Abel, N. (2001). From metaphor to measurement: Resilience of what to what? *Ecosystems, 4*, 765–781.

Dasgupta, P., & Mäler, K.-G. (2000). Net national product, wealth and social well-being. *Environment and Development Economics, 5*, 69–93.

Dasgupta, P., & Mäler, K.-G. (2001). Wealth as a criterion for sustainable development. *World Economics, 2*, 19–44.

Dasgupta, P., & Mäler, K.-G. (2003). "Intertemporal welfare economics in imperfect economics," Beijer Discussion paper 140. Stockholm, Sweden: Beijer Institute.

Folke, C., et al. (2002). Resilience and sustainable development. *Ambio, 31*, 437–440.

Gunderson, L. H. (2000). Ecological resilience – in theory and application. *Annual Review of Ecology and Systematics, 31*, 425–439.

Gunderson, L. H., & Holling, C. S. (Eds.) (2001). *Panarchy: Understanding transformation in human, and natural systems*. Washington, DC: Island Press.

Heal, G. M. (1998). *Valuing the future: Economic theory and sustainability*. New York: Columbia University Press.

Levin, S., et al. (1998). Resilience in natural and socioeconomic systems. *Environment and Development Economics, 3*, 222–235.

Ludwig, D., Carpenter, S. R., & Brock, W. A. (2003). Optimal phosphorus loading for a potentially eutrophic lake. *Ecological Applications, 13*, 1135–1152.

Mäler, K.-G., Xepapadeas, A., & de Zeeuw, A. (2003). The economics of shallow lakes. *Environmental and Resource Economics, 26*(4), 105–126.

National Academy of Sciences (1999). Nature's numbers: Expanding the national economic accounts to include the environment. Washington, DC: National Academy of Sciences.

National Academy of Sciences (2000). *Our Common Journey*. Washington, DC: National Academy of Sciences.

Postel, S., & Carpenter, S. R. (1997). Freshwater ecosystem services. In G. Daily (Ed.), *Nature's services* (pp. 195–214). Washington DC: Island Press.

Reed, T., & Carpenter, S. R. (2002). Comparisons of P yield, riparian buffer strips, and land cover in six agricultural watersheds. *Ecosystems, 5*, 568–577.

Scheffer, M., Carpenter, S., Foley, J., Folke, C., & Walker, B. (2001). Stochastic events can trigger large state shifts in ecosystems with reduced resilience. *Nature, 413*, 591–596.

Swiss Re. (1998). Environmental management systems and environmental impairment liability insurance. Zurich, Switzerland.

World Bank. (1997). Expanding the measures of wealth. Washington, DC.

Xepapadeas, A., (forthcoming). Economic growth and the environment. In Arrow. K., & Intrilligator, M. D. (Series Ed.) and Mäler, K.-G., & Vincent, J. (Volume Ed.), *Handbook of Environmental Economics* in the series *Handbooks of Economics*. Amsterdam, the Netherlands: Elsevier.

Chapter 32
Askö 2001: Commentary by Lance Gunderson

Lance Gunderson

'Sustainability's compass: indicators of genuine wealth' is a well crafted, well reasoned and well written article that provides thoughtful and grounded guidance for seeking sustainable futures. This article is but one of a long series of such products that arose from the simple idea of bringing together the world's best thinkers from relevant ecological and social sciences (primarily economics) to seek solutions for some of our most vexing or 'wicked' problems. The article deftly weaves together core concepts from economics, systems ecology and one from the interdisciplinary field of ecological economics. This article is among the few works (e.g. Carpenter et al. 1999; Carpenter and Brock 2004; Ludwig et al. 2005 among others) that integrates economic and ecologic models for confronting and managing dynamic complexity in non-linear, resource systems.

The authors correctly focus on resilience, the key ecological property that mediates the transition among multiple, alternative ecosystem regimes. The word resilience originated from the Latin roots 'to jump or leap back'. Holling (1973) described resilience as the capacity of an ecological system (such as a forest) to absorb disturbances (such as storms, fire or pests) and retain similar characteristics and as the amount of disturbance that a system can absorb without undergoing a change in structure and function. The authors provide one of the best understood and rigorous explanations of resilience, using lake water quality.

The authors focus on development of system indicators to signal system resilience and system state. They also argue of the need for these indicators to track various forms of capital that contribute to resilience. Thus, a variety of forward-looking (market) mechanisms can be created to manage capital (especially natural capital) needed to maintain ecological resilience. Throughout the article, the authors focus on the need to maintain a single ecological state through market and other interventions. That is, management should focus on keeping the system within bounds or thresholds of a single beneficial and desired regime. Yet, the world is more complicated, with more variations than maintaining a single desired regime.

L. Gunderson
Department of Environmental Studies, Emory University, USA
e-mail: lgunder@emory.edu

T. Söderqvist et al. (eds.), *Bringing Ecologists and Economists Together*,
DOI 10.1007/978-90-481-9476-6_32,

Multiple or alternative ecological domains can be viewed and valued in a range of ways by society. Simply, these multiple domains can be viewed as positive, negative or neutral in a normative sense. The authors focus on the role of resilience in maintaining a desired ecological state (positive resilience). That is, the authors argue (correctly) of the need for mechanisms to protect against the loss of ecological resilience and change of state and its corollary flow of ecosystem goods and services and economic activities. Yet, many examples exist where a) it may not matter to the social or economic systems as to the ecological state, or b) resilience may be a negative attribute (e.g., the system is in a social trap) or c) the system has already flipped to an alternative state, and people have adapted to the new regime. For example, wool production in New South Wales, Australia, has been stable for over a century, in spite of a change of ecological state in the rangeland (Abel et al. 2000). In the wheat belt of western Australia, the socio–ecologic system is very resilient to change, exhibiting a 'lock-in' trap (Allison and Hobbs 2004), from which indicators or forward looking market mechanisms have not helped escape. Finally, many of the largest conservation and environmental issues of our time occur in places where the system is already in an undesired domain. In the Everglades of the United States, endangered species populations and vegetation conversions are both examples where the system has already crossed a threshold into an undesired state (Gunderson 2001). In these cases, infusions of capital (human, fiscal, social and political) are needed to help restore or recover from a degraded state (Olsson et al. 2006).

The issue of regime management is central to sustainability. The authors have made a valuable contribution to sustainability science by linking the management and maintenance of key forms of capital with indicators of resilience and forward looking economic and institutional structures. These are applicable to some of the other situations raised in the previous paragraph. Integrating ecologic, economic and other social dimensions is a predicate for resolution of chronic and problematic resource management issues. Creating alternative futures, followed by exploration and navigation of paths taken and not taken, will help build adaptive capacity. It is our ability to learn and confront uncertainties of the dynamics of coupled systems of humans and nature that will help navigate a path to sustainability.

References

Abel, N., Ive, J., Langston, A., Tatnell, B., Tongway, D., Walker, B., et al. (2000). Resilience of NSW rangelands: A framework for analysing a complex adaptive system. In P. Hale, A. Petrie, D. Moloney, & P. Sattler (Eds.), *Management for sustainable ecosystems* (pp. 59–71). Brisbane: The Centre for Conservation Biology.

Allison, H. E., & Hobbs, R. J. (2004). Resilience, adaptive capacity, and the "lock-in trap" of the Western Australian agricultural region. *Ecology and Society, 9*(1), 3. [online] URL: http://www.ecologyandsociety.org/vol9/iss1/art3/.

Carpenter, S. R., & Brock, W. A. (2004). Spatial complexity, resilience and policy diversity: Fishing on lake-rich landscapes. *Ecology and Society, 9*(1), 8. [online] URL: http://www.ecologyandsociety.org/vol9/iss1/art8.

Carpenter, S., Brock, W., Hanson, P. (1999). Ecological and social dynamics in simple models of ecosystem management. *Conservation Ecology, 3*(2): 4. [online] URL: http://www.consecol.org/vol3/iss2/art4/.
Gunderson, L. H. (2001). Managing surprising ecosystems in southern Florida. *Ecological Economics, 37*, 371–378.
Holling, C. S. (1973). Resilience and stability of ecological systems. *Annual Review of Ecology and Systematics, 4*, 1–24.
Ludwig, D., Brock, W. A., & Carpenter, S. R. (2005). Uncertainty in discount models and environmental accounting. *Ecology and Society, 10*(2), 13. [online] URL: http://www.ecologyandsociety.org/vol10/iss2/art13/.
Olsson, P., Gunderson, L. H., Carpenter, S. R., Ryan, P., Lebel, L., Folke, C., et al. (2006). Shooting the rapids: Navigating transitions to adaptive governance of social-ecological systems. *Ecology and Society, 11*(1), 18. [online] http://www.ecologyandsociety.org/vol11/iss1/art18/.

Chapter 33
Askö 2001: Commentary by Scott Barrett

Scott Barrett

In what direction ought a society to head? Does there exist a compass – a measure of some kind – that can tell us whether society is advancing in that direction? Is there a way to know if society is badly off course? These questions are at the heart of the paper "Sustainability's compass: indicators of genuine wealth." They are among the most important questions a society can ask.

The measure proposed in the paper is "genuine wealth." The paper does not directly address the question of the direction in which a society ought to head, but if "sustainability" is taken to be a priority, and if the feasibility constraints allow, then genuine wealth ought not to decline. A country whose genuine wealth is falling, and showing no signs of turning around even when this is feasible – such a country, it may safely be said, is badly off course. Table 31.1 in the paper tells us that, by this reasoning, there are a number of countries heading in the wrong direction.

Ecologists emphasize the importance of ecosystem resilience, whereas economists think in terms of trade-offs – an important one being the trade-off between the present and the future (thus the importance of "forward-looking indicators"). The paper unifies these concerns, while also distinguishing between them. In the example of the lake, for example, an increase in the concentration of phosphorus (P) in the sediments is taken to be a warning sign of possible future eutrophication. The accounting price for P, if calculated correctly, would reflect this vulnerability. As the paper suggests, this price can be thought of as the premium a society is willing to pay to insure against a loss in ecosystem services caused by eutrophication.

As explained in the paper, in a sense, either measure will do, provided we focus only on the lake. The ecologist's P and the economist's accounting price are fully compatible as indicators of the quality of the lake.

When addressing a broader set of challenges, this consonance of perspective may be harder to sustain. Genuine savings needs to consider more assets than the ecological integrity of a lake, and that means that accounting prices must reflect another kind of tradeoff – the necessity of having less of one thing in order to have

S. Barrett
School of International and Public Affairs & Earth Institute, Columbia University, USA
e-mail: sb3116@columbia.edu

T. Söderqvist et al. (eds.), *Bringing Ecologists and Economists Together*,

DOI 10.1007/978-90-481-9476-6_33,

more of something else ("substitutability"). The paper alludes to this problem, but does not face it squarely. It says that, "in dealing with ecological wealth, one has to accept conditions of limited substitutability, or thresholds above which substitution is not possible"(p. 5). It then says that, until these kinds of tradeoffs are understood, "a variety of indicators should be employed that represent diverse aspects of ecosystem wealth" (p. 5). I take this to mean that the calculation that the paper makes in the case of the lake should be repeated for other ecosystems. This is certainly to be desired. But decisions have to be made – and are being made every day – that affect ecosystems and societal wellbeing. Uncomfortable tradeoffs need to be confronted.

Let me illustrate what I mean by considering a different kind of problem: global climate change.

The possibility of "abrupt" climate change is analogous to a lake moving to a eutrophic state. The slow moving variable in this case is atmospheric concentrations of greenhouse gases. Higher concentrations make abrupt climate change more likely, implying that a premium needs to be added to the accounting price on "gradual warming" – a premium on incremental changes in the concentration level that increases as the concentration level increases.

This situation is similar to that of the lake, but in this case tradeoffs cannot be set to one side. A decision has to be made between reducing emissions today and carrying out more R&D in order to make it easier to reduce emissions in the future. A decision has to be made between reducing emissions today and carrying out R&D into a malaria vaccine. (Reducing emissions today will reduce the spread of malaria in the future, but a malaria vaccine would reduce the consequences of malaria in the future *in general*, and not only in the areas where malaria is likely to become more prevalent because of climate change.) A decision has to be made between reducing emissions today and distributing mosquito-proof bednets today. (The former would reduce the spread of malaria in the future. The latter would reduce the incidence of malaria today.) A decision has to be made between reducing emissions today by a little, by means of conservation, or by a lot, by substituting nuclear power for fossil-fuel-generated electricity. (Reducing the risk of abrupt climate change significantly would increase the risk of a nuclear incident.) All of these choices are unavoidable. Indeed, they are being made right now.

The example of climate change exposes another challenge to the measurement of genuine savings. The paper asks, "At what spatial scale should ecological wealth be assessed – nations, major river basins, the world?" (p. 4) To ecologists, the unit of analysis is the ecosystem: the lake, the forest, the ocean – and, in the case of climate change, the Earth itself. The concept of genuine savings, as it is usually presented, sees things differently. It assumes, or presumes, that the unit of analysis is the state. The estimates in Table 31.1 are for countries, not ecosystems.

There is a good reason for this. Genuine wealth depends on the endowments that are given (including the states of various ecosystems) and the choices that are made about how these endowments are used. These choices are made by individuals and various kinds of social organization, including families, communities, and corporations. *Their* choices, however, are shaped by the state. States alone have the power to tax, to quarantine, to conscript, to regulate – interventions that help determine genuine wealth.

A problem arises when the jurisdiction that is appropriate for ecosystem analysis differs from that for governance. The World Bank, in its own calculations of genuine wealth, puts an accounting price on greenhouse gas emissions that reflects the *global* consequences of those emissions. This approach makes sense if the unit of analysis is the world as a whole. But when applied to individual countries, as it is in the World Bank's calculations, I'm not convinced it is appropriate, for it ignores distributional considerations – aspects that are important in their own right, and that would also prove important in international negotiations, even successful ones that sustained a globally optimal outcome.

The two greatest achievements of this kind to date are the protection of the ozone layer and the eradication of smallpox. In both cases, every country contributed to a universally desired outcome. But in both cases, rich countries compensated poor countries for taking the actions needed to supply these global public goods. The world has so far been unsuccessful in addressing the threat of global climate change. However, under ideal conditions, including full participation and frictionless application of all the flexible mechanisms in the Kyoto Protocol, this agreement would equalize the marginal costs of abating greenhouse gas emissions everywhere, but with the rich countries paying for the abatement undertaken in poor countries. The calculations by the World Bank do not make allowance for such transfers. The transfers would not matter for a global measure of genuine wealth, but should they be excluded from calculations of genuine wealth for individual countries?

This discussion is only to agree with the paper, not only in the things it positively asserts, but also in the questions it asks and leaves unanswered. As the paper says, more research on this topic is needed. It should keep the Beijer Institute busy for a few more years yet.

Chapter 34
Askö 2002: Coping with Uncertainty – A Call for a New Science-Policy Forum*

Ann P. Kinzig, David Starrett, Kenneth Arrow, Sara Aniyar, Bert Bolin, Partha Dasgupta, Paul R. Ehrlich, Carl Folke, Michael Hanemann, Geoff Heal, Michael Hoel, AnnMari Jansson, Bengt-Owe Jansson, Nils Kautsky, Simon Levin, Jane Lubchenco, Karl-Göran Mäler, Stephen W. Pacala, Stephen H. Schneider, Domenico Siniscalco, and Brian Walker

---

*This article was published in *Ambio,* Vol. 32, Kinzig et al., Coping with uncertainty: a call for a new science-policy forum, 330–335, Copyright the Royal Swedish Academy of Sciences (2003).

A.P. Kinzig
School of Life Sciences, PO Box 874501, Arizona State University, Tempe, AZ 85287-4501, USA

D. Starrett
PO Box 150821 Ely, NV 89315, USA

K. Arrow
Department of Economics, Stanford University, USA

S. Aniyar, C. Folke and K.-G. Mäler
Beijer Institute of Ecological Economics, the Royal Swedish Academy of Sciences, Stockholm, Sweden

B. Bolin
Department of Meteorology, Stockholm University, Sweden

P. Dasgupta
Faculty of Economics, University of Cambridge, UK

P.R. Ehrlich and S.H. Schneider
Center for Conservation Biology, Stanford University, USA

M. Hanemann
Department of Agricultural and Resource Economics, University of California, Berkeley, California, USA

G. Heal
Program on Information and Resources, Columbia University, New York, USA

M. Hoel
Department of Economics, University of Oslo, Norway

A. Jansson, B.-O. Jansson, and N. Kautsky
Department of Systems Ecology, Stockholm University, Sweden

S. Levin and S.W. Pacala
Department of Ecology and Evolutionary Biology, Princeton University, USA

J. Lubchenco
Department of Zoology, Oregon State University, USA

D. Siniscalco
Fondazione Eni Enrico Mattei, Milan, Italy

B. Walker
CSIRO Sustainable Ecosystems, Canberra, Australia

T. Söderqvist et al. (eds.), *Bringing Ecologists and Economists Together*,
DOI 10.1007/978-90-481-9476-6_34, © Springer Science+Business Media B.V. 2011

The scientific and policy worlds have different goals, which can lead to different standards for what constitutes "proof" of a change or phenomena, and different approaches for characterizing and conveying uncertainty and risk. These differences can compromise effective communication among scientists, policymakers, and the public, and constrain the types of socially compelling questions scientists are willing to address. In this paper, we review a set of approaches for dealing with uncertainty, and illustrate some of the errors that arise when science and policy fail to coordinate correctly. We offer a set of recommendations, including restructuring of science curricula and establishment of science-policy forums populated by leaders in both arenas, and specifically constituted to address problems of uncertainty.

34.1 Introduction

Throughout their history, humans have devised a number of approaches and strategies to reducing (perceived) uncertainty or the psychological impacts of uncertainty, and to function and thrive in a partially unknown world. A certain world begets certain 'best strategies' for survival; an uncertain world necessitates a larger portfolio of options and approaches, and means for choosing among them.

We cannot hope to cover the whole of human history here. Instead, we focus on the ways in which uncertainty and approaches to reducing uncertainty have played out in recent history in two arenas; science and policy. We focus on these because it is the interplay between these that has relevance for global environmental futures, and because the characterizations of uncertainty, and strategies for coping with it, differ so dramatically between the two.

The scientific process is built on the goal of advancing knowledge; penetrating and reducing the reaches of what is not known. Each advance is built on knowledge acquired earlier. The cost of 'incorrect knowledge' is therefore quite high, affecting not only that building block in the foundation but those that follow. Science has therefore evolved procedures whose primary goals are to protect against being wrong.

The policy process is built on the goal of addressing societal ills or challenges. Timeliness is frequently of the essence; action may have to precede knowledge. The errors that need be avoided are those associated with undue political or social costs, including harm to constituents, the economy, national security, or the environment.

Because of these fundamentally different goals, science and policy have evolved significantly different 'evidentiary standards'. Scientists apply relatively high evidentiary standards – insisting, for instance, that there be only a small probability (sometimes 5% but frequently lower than 1%) that an incorrect conclusion has been drawn. This means that many conclusions concerning the workings of the world that have a reasonable probability of being correct – say 75% – may not pass muster.

The policy process employs both stricter and looser evidentiary standards, largely based on the (perceived) costs of being wrong. If we are certain a patient is going to die shortly, for instance, there is little hazard in prescribing a drug whose efficacy is largely unknown, but offers some hope of life extension. Here the unknown interacts with our perceptions of risk in such a way as to allow us to employ relatively loose evidentiary standards. A drastically different evidentiary standard should apply for instance, in the case of cosmetics, where society insists on a vanishingly small chance that one might die from the application of mascara. The manufacturer must demonstrate the highest confidence that the product works as intended and is very unlikely to have undesirable side effects.

Difficulties arise due to these different standards in four ways:

(i) Failing to recognize that different standards in policy and science exist; this can result in a failure to communicate properly. Policymakers may believe that scientists are using a stricter or, more likely, looser evidentiary standard than is actually being employed. Thus, an outcome that a scientist believes is highly certain may be (mis)interpreted as only probable or likely in policy circles.

(ii) Difficulty arises because scientists may fail to provide information that could be considered useful or even crucial in the policy arena, in large part because they may have failed to study the appropriate phenomena. The success of scientists depends on their ability to advance their field under the relatively strict evidentiary standards that apply. This selects for fairly reductionist approaches to studying phenomena whose drivers can be tightly controlled and manipulated. Until recently, this meant largely avoiding study of precisely the kinds of environmental systems that society is influencing most profoundly – global climate systems, entire ecosystems or landscapes. Fortunately, our capacity to study these complex, and often highly uncertain, systems has advanced tremendously, but the strong scientific tradition of applying strict evidentiary standards can still delay the introduction of information that could be considered crucial by policymakers.

(iii) Scientists may find it difficult to quantify the uncertainty associated with highly uncertain, and to some extent unknowable, futures. This has led, in some circles, to an avoidance of characterizing uncertainty at all. Witness, for example, the recent emissions scenarios of the IPCC, each of which was presented without an assessment of likelihood, and each of which was thus treated as equally likely by the public and policymakers. No rational assessment of probabilities of the various scenarios would conclude equal likelihood; rather such a process would conclude (potentially drastically) different likelihoods for each scenario, albeit with very low confidence (Schneider 2002). We will discuss the need for likelihood assessment and relevant methodology in the next section.

(iv) The neutral language of evidentiary standards can be used to mask what is ultimately a debate over values. This tactic is not limited to the (supposedly subjective) policymaker, but is employed by the (supposedly objective) scientist as well. Most if not all of the environmental problems faced by society are complex all the way down. They are rife with uncertainty and there can be numerous plausible solutions, leading to numerous possible futures. These futures will differ in terms of environmental protection, social justice, economic growth, and political freedoms, among other things. Scientists can help illuminate those futures and trade-offs, but they are no more expert than any other citizen when choosing among the trade-offs. There is nothing objective about valuing environmental protection over political freedom, or economic growth over social justice, and no set of evidentiary standards can ultimately allow society to navigate among these complex considerations. Unfortunately, they are often employed as if they can. Scientists advocate certain policies, frequently claiming a scientific objectivity when they do, but their positions ultimately derive from their own values. Policymakers invoke scientific uncertainty as the justification for inaction on certain issues, when ultimately their positions have more to do with the weight of economic and environmental considerations. In what follows, we will attempt to sort out the roles of objective versus subjective information and values (Funtowicz and Ravetz 1991).

34.2 Principles and Examples

How should scientists present information to the public and to decision makers? In particular, how should information be presented when there are substantial uncertainties concerning what is known – or even knowable? In discussing these questions, we shall invoke two well-established principles: *Expected Utility theory* for making decisions at a point of time in the face of uncertainty, and *Bayesian updating* for revising our measures of uncertainty in the presence of new information. According to expected utility theory, for any action we might take we must first identify each of the possible outcomes (consequences) it might engender and determine the relative likelihood of each consequence. We then assign utilities to consequences in a way that reflects relative desirability of outcomes and our attitudes toward risk. We then take the action that maximizes expected utility. Probabilities are determined using the best available information. When new information becomes available we use it to update employing Bayes' rule. If the new information makes an outcome more likely to be observed than it would have seemed given *ex ante* information, the probability of that outcome is increased using the formulas of conditional probability, and *vice versa* (Morgan and Henrion 1990).

Neither of these principles is particularly intuitive to the average citizen but there is a long literature defending them against plausible alternatives.[1] We will not survey that literature here although some of the supporting arguments will be implicit in our treatment of the subject.

[1]For a relative nontechnical introduction to the theory of choice under uncertainty, see Raiffa (1968). For an introduction to Bayesian methods, see Lee (1997). More detailed and comprehensive treatments of these subjects can be found in Berger (1985), Pratt et al. (1995), and Press (2003).

34.2.1 Expected Utility

To illustrate the principle of expected utility and show how it impacts the way in which scientists should conduct and report their work, let us see how it requires us to optimally trade-off type I versus type II error. Consider a stylized example involving whether or not to do a mastectomy. Suppose there are two states of the world, (a) in which the outcome generates a higher utility U(a) than doing nothing, and (b) in which it generates lower utility U(b). Normalizing so that the *status quo* gets zero utility, we have $U(a) > 0 > U(b) = -C(b)$ where C(b) represents the costs, both monetary and psychological, of performing an unnecessary or undesirable procedure. Assume further that we know the probabilities of these states: P(a), $P(b) = 1-P(a)$. If we do the procedure we incur the danger of type I error (accepting the procedure when it is unjustified) which would happen with probability P(b). But, if we forgo the procedure we experience type II error with probability P(a).

According to the principle of expected utility we should conduct the mastectomy if and only if the *expected* utility of proceeding exceeds zero (the expected utility of doing nothing). Here the rule would be to proceed when $P(a)U(a) > P(b)C(b)$. Observe that this rule trades off the two types of error, taking into account the benefits of being right relative to the costs of being wrong. In this regard, it is extremely important that scientists and practitioners see the rewards and benefits in the same way as the society being served. For example, if doctors have an inordinate fear of malpractice claims while the medical costs of procedures are covered by insurance, they may be induced to conduct procedures unless P(a) is extremely small (P(b) is extremely large), in order to protect themselves. Thus, they may conduct even when $P(a)U(a) < P(b)C(b)$. In the extreme they would see only the benefits of desirable procedures and not the costs of unnecessary ones, and would conduct any time that $P(a) > 0$. Clearly society, which must incur these costs, loses on balance.

In other areas of science we may see a bias in the opposite direction. As indicated earlier, scientists tend to impose high evidentiary standards, and frequently will not report results or recommend action unless P(a) is very large, e.g. 95 (P(b) is very small). Policy makers, relying on this may then fail to act when $P(a) < 0.95$, yet $P(a)U(a) > P(b)C(b)$, and we incur the reverse bias. Although it is probably too early to tell now, inaction on the consequences of global warming may be attributable to failures of scientists to take strong enough positions for fear of possibly being wrong. Indeed, this example suggests that reliance on any arbitrary "confidence level" is inimical to good decision-making; rather scientists should report results with whatever levels of confidence can be justified.

Of course, scientists must still impose a severe level of standard before accepting a new theory into the corpus of scientific knowledge due to the substantial costs associated with building on a false scientific base. Thus, all practitioners should get used to the idea that different standards are appropriate depending on context.

34.2.2 Subjective Probability

Perhaps the most controversial aspect of our operative principles involves the use of subjective probability. The concept of probability is easiest to explain and defend in situations where repeated trials can be performed. For example, most would agree that a die is fair if after a large number of repeated rolls, each of the six numbers comes up approximately one-sixth of the time. We are then comfortable in saying that the likelihood or *probability* of any particular number on a given throw is one sixth. In this regard, the insurance industry uses data on frequencies of illness, death, accident, property damage, etc. within a given demographic category to determine the likelihood that any individual in that category will be affected, and offers protection against the bad outcome using these numbers to determine premiums. Given their ability to spread risk over many people, they can offer coverage at nearly "fair" odds (based on the frequencies), and risk–averse individuals find it advantageous to purchase coverage. This sort of inference and use of probability from frequency is well established and not very controversial.

However, when repeated trials are infeasible, as they often are in real world situations, it is not clear how probabilities could or should be assigned and used. For example, consider the event that nuclear bomb-making material falls into the hands of terrorists. Since the opportunity has only been around for a relatively short time and this outcome has never been documented, there is no way to infer likelihood from the principle of repeated trials. Yet, expected utility theory would require that we assign a "best guess" probability to this event.

The logical underpinnings of subjective probability as applied to decision-making are similar to the logic of value comparisons in welfare economics. Economists have long argued that one must consider tradeoff rates between the value of marketed goods – as measured in money units – and such intangibles as, e.g. a statistical life. This comparison is necessitated by the fact that policy makers must make decisions on projects that have some monetary benefits but incur some risk of loss of life. Whatever decision is made clearly will have some implications concerning the implicit tradeoff used, and consistency across decisions dictates that the same trade-off rate be used whenever the conflict arises. This trade-off rate then defines the "shadow price" of a statistical life. Similar logic dictates the use of subjective probabilities. For example, suppose we are considering employing a breeder reactor whose dollar net benefits have been measured as USD 10 billion. We believe, however that its use will increase the possibility that bomb-making material falls into the hands of terrorists. Should that occur, we anticipate substantial costs; assume for sake of argument that we can measure these and they come out to USD 1 trillion. Still a decision cannot be made without some judgment concerning likelihood and aversion to risk. Concerning risk, assume either risk neutrality or that the numbers above are already in utility units. Then if we go ahead with the project we are implicitly saying that the likelihood is increased by less than one chance in 100, whereas if we forgo we are making the reverse judgment. Again, consistency across decisions requires that we use the same number in all similar comparisons and this number is what we call the subjective probability.

There are many difficulties in measuring subjective probabilities, and again these parallel similar difficulties in measuring shadow prices. In some instances, we may be able to learn something about relevant citizen's attitudes from observed behavior. For example, people's willingness to build bomb shelters during the cold war tells us something about the perceived likelihood of a nuclear exchange. Or the wage premium on life-threatening jobs should tell us something about the value of a statistical life. Unfortunately, such "hedonic" methods rarely if ever allow us to exactly identify a shadow price or subjective probability (Starrett 1988), and in any case can only reflect the quality of information possessed by citizens. The bottom line is that decision makers still need to use best judgment without a firm objective basis in specifying shadow prices and subjective probabilities. Scientists therefore have an obligation to provide the highest possible quality of information concerning relative likelihood of events and values of intangibles (Moss and Schneider 2000).

34.2.3 Bayesian Updating

The main way in which scientists can inform concerning probabilities – aside from repeated trials when the frequency method is available – is through the conducting of controlled experiments and tests. In the area of health, the questions might be on whether or not large doses of vitamin C helps prevent the common cold or whether a glass of red wine per day reduces the risk of heart disease. Evidence is brought to bear by comparing the experience of groups who take the treatment with a control group of those who do not. According to Bayes' law, what we learn from such evidence depends on three things: (i) the test results; (ii) accumulated past evidence; and (iii) the inherent accuracy of such tests. Suppose that a new test does suggest a positive relationship between red wine and reduced heart disease. This evidence will be most compelling in changing our estimates of likelihood if there is little previous evidence to go on and we have reason to believe that such tests are accurate. Unfortunately, there is something of a conflict here, because if we have little previous experience with such controlled studies, we have no way to judge their inherent accuracy.

Consider now another situation where there is considerable accumulated evidence, but it is inconclusive in that some studies point one way and some the other. One might think in this circumstance that a new study would be quite informative, but this is not generally true. The fact that previous studies of this sort have been inconclusive suggests that the method is not very accurate, and adding one new study showing a positive relationship should not change our prior likelihood of a relationship very much. We suggest that this situation is fairly typical of the medical literature on nutrition where conflicting studies abound. In such circumstances, medical researchers do a disservice when they make broad claims on the basis of one new study. For one thing, such a study should not alter our priors much; but more importantly, when such claims turn out to be unjustified, as they sometimes do, the public loses confidence in the general efficacy of medical research and tends to ignore warnings that are fully justified on the basis of accumulated evidence.

34.2.4 Functional Uncertainty and Surprises

Of the many types of uncertainty, among the least recognized, understood, and dealt with is something we will call *functional* uncertainty. We may be ignorant concerning the relevant state space or more generally with the functional relationships involved in the policy issue at hand. There are no general principles to invoke here so we will simply illustrate with two examples. (i) Think about the Georges Bank fishery. Marine economists and ecologists early worked with a logistics model in which each positive fish stock would be associated with a positive sustainable level of fishing. Only later did evidence accumulate that this probably was the wrong model and that a correct version might entail a threshold whereby once the stock is allowed to fall to a critical level, the stock would collapse even if all fishing was curtailed. The fishery would be left barren, with quite dire consequences for the fishing industry. (ii) Consider the North Atlantic Thermohaline Circulation, wherein Gulf Stream warm water is channeled north in exchange for a cold water return from arctic regions. We are quite uncertain exactly how much disturbance – in the form of temperature or precipitation change – could trigger an instability, although we do know such instability exists, just not where the thresholds are (Mastrandrea and Schneider 2001). Without this knowledge, we cannot predict confidently when human-made forces might shut it off. We will return to this example in our discussion of irreversibility below.

In each of these cases, functional uncertainty introduces the possibility of a "surprise," and such surprises have been a recurring feature of science throughout its history. How should scientists and policy makers deal with the possibility of surprises? Obviously, by their very nature surprises cannot be fully anticipated, yet they can have far-reaching consequences. One strategy for dealing with surprises would be *avoidance*: namely do whatever we can to minimize the occurrence of a surprise. But even if desirable, this strategy would have the fatal flaw of being non-operational. Since we can never know for sure what will trigger a surprise, we cannot know which strategies minimize the trigger. *Ex post*, of course, we may observe the trigger and identify the desired strategy. For example, knowing what we know now, we realize that fishing should have been restricted earlier on the Georges Bank (Holmes 1994). However, *ex ante* it was not clear when or by how much this should be done. Indeed, it is even possible that allowing stocks to over-produce could promote a pathogen that provides another equally destructive surprise.

Further, we would argue that even if the avoidance strategy could be made operational, it would not be desirable. Suppose, for example we knew that due to natural ecological uncertainty, there was some small probability that a fishery would collapse in any given year, and moreover that this likelihood was an increasing function of the level of fishing. Then following the avoidance strategy would require that we never fish. But such a strategy would violate the principle of tradeoff that underlies all good decision-making: The likelihood of a "bad" surprise must be traded off against the lost benefit (economic and social) from fishing. Again, we cannot avoid making a subjective judgment concerning the likelihood of surprise and using it to compare the risks against lost benefit. More generally, extreme strategies such as "always assume the worst (or the best)" are never desirable because they ignore relevant trade-offs.

There is an emerging literature on alternative strategies to use in managing surprise, including strategies to improve anticipation of surprise and to cope with inevitable surprise (Schneider et al. 1998; Gunderson and Holling 2002; Berkes et al. 2002).

This should not be taken, however, as a necessary argument for continuation of *status quo* practices. The point is to weigh relevant trade-offs. There are numerous examples of overfishing of marine resources (Jackson et al. 2001) and there is much to learn from what has happened in Georges Bank and other areas. In the Baltic Sea, for example, the cod spawning stock is outside of its biological limits (HELCOM 2002) in spite of many years of scientific advice to reduce catches. Annual assessments of the declining cod spawning stock and full awareness of the fact that spawning success depends on the unpredictable inflows of salty, oxygenated water from the North Sea – a "standing surprise" – did not persuade the decision-makers to adopt precautionary practices, though in this case they should have. Adequate information was there, but fierce lobbying by fisheries organizations and the difficulties of obtaining international consensus prevented beneficial outcomes (Fig. 34.1).

34.2.5 Surprises and the Strategy of Scientific Pursuit

How much should scientists "fool" with Mother Nature? Should we engage in genetic engineering? Should we try to interfere with the dynamics of hurricanes? Should we build new types of nuclear reactors? Clearly, any of these strategies take us into unknown places where the possibility of surprises increases. But again there are trade-offs. For one thing, such projects, by taking us outside known parameters, provide information for resolving some functional uncertainty. Genetic experimentation, for example, can tell us things about the chemistry of life that we might not be able to discover otherwise. For another, some surprises are good; indeed many of our most

The total fish catch in the Baltic increased dramatically in the twentieth century up until about 1985.
Since then, it has declined equally dramatically.
(Photo: G. Aneer)

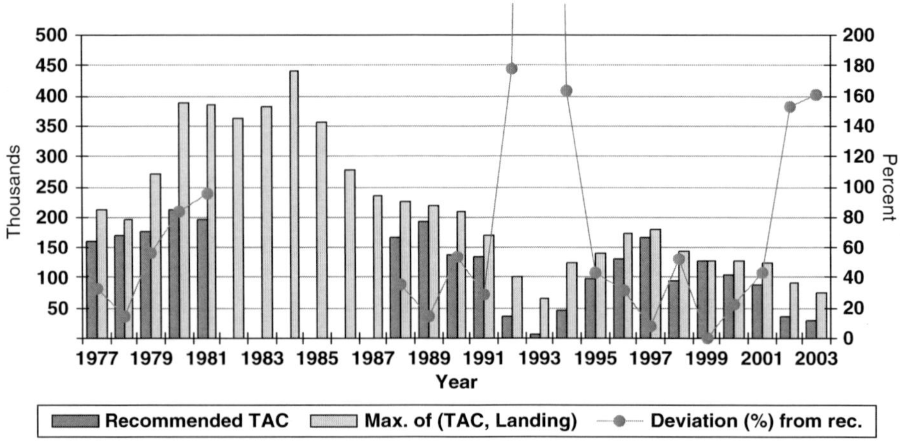

Fig. 34.1 Recommended Total Allowable Catches and the outcomes of subsequent management actions. (IBSFC = International Baltic Sea Fishery Commission). The Management Agency (IBSFC) agrees on one TAC for cod stocks and the whole Baltic and allocates it to its members. Each country is responsible for the enforcement of its national quotum. The maximum of agreed TAC, and reported landings are shown here together with percent deviations from the recommendations. The agreed TACs for the years 1979–1981 were not allocated between nations, no agreement was reached for the years 1982–1988. The reported catches for the years 1992–1995 and 2000–2002 are of highly uncertain nature due to misreporting. In general, we can see that periods in which reported catch exceeds recommendations were followed by decline in fishery output, suggesting that in this instance we had enough information to correctly regulate the fishery, but lacked the political will. The authors would like to thank Dr. Bengt Sjöstrand at the Institute of Marine Research, Lysekil, Sweden, for compiling and analyzing the data and for providing this figure

important scientific discoveries, e.g. x-rays, have come as surprises observed because a scientist did something that had not been done before and consequently observed something unexpected. This conflict between maximizing the opportunity for learning and minimizing the likelihood of "bad" surprises is fundamental to the scientific endeavor, yet has no easy resolution.

34.2.6 Attitudes toward Risk

One of the things the general public expects from science is the reduction or elimination of uncertainty faced. It is clear to us now that uncertainty cannot be eliminated entirely, but there remains the question as to how much in the way of resources we should devote to reducing uncertainty and how the presence of residual uncertainty should be incorporated in our decision-making rules.

We can measure the aversion to residual uncertainty by asking how much a person who is facing some random outcome (for example chance of traffic accident or property damage) would be willing to pay to obtain an outcome with certainty that has the same expected value. In most circumstances people are willing to pay something (which we refer to as the *risk premium*) and that explains how the insurance companies are able to make money by spreading the risk. Risk aversion is incorporated in decision-making by using a utility function which is nonlinear (in fact concave) on numeraire (monetary) units. Concavity means that we object more to downside losses than we are happy with upside gains, and the more concave our function, the stronger this effect, and the greater the degree of risk–aversion represented (Box 34.1).

What degree of risk–aversion should be employed in public decision-making? It should be obvious that the answer depends on context. For example, in cases where risks are socially insurable we should provide the insurance whatever else is done and treat the residual problem in a risk–neutral way. However, few risks are completely

Box 34.1 Determination of the Risk Premium

Consider a worker in a highly risky job. For illustrative purposes suppose half the time she will make USD50K and half the time nothing (so that her expected income is USD 25K). We can determine the degree of aversion to risk by asking how much money the worker would accept for certain rather than have to face the job risk. A worker who will accept less than USD 25K is said to be *risk–averse*, and the difference between USD 25K and what she will accept is labeled the risk premium. In expected utility theory, risk aversion is captured by use of a utility function which is concave in dollars. If we normalize so that U(50K) = 1 and U(0) = 0, such a function looks as in the diagram below, with the expected utility of the risky job equal to 0.5. As drawn, the worker has said she would accept USD 7.7K for sure (U(7.7K) = 0.5), implying a risk premium of USD 17.3K. Note that if the worker were risk–neutral (no risk premium) the implied utility function would be linear and further that the degree of risk aversion will be positively related to the degree of concavity in the utility function.

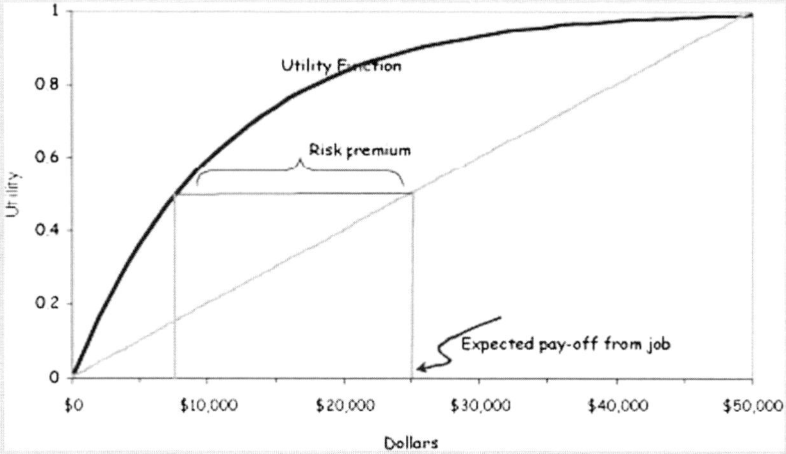

insurable (whereas we can insure against the monetary costs of illness or accident, we cannot insure against the psychological costs) and some are not insurable at all. The latter include events that affect the social group as a whole such as ozone depletion or an asteroid impact, as well as the functional uncertainties discussed earlier.

When uninsurable risks are involved, we must add the degree of risk aversion to subjective probabilities and shadow values as things that need to be specified as a basis for good decision-making. The degree of risk aversion not only serves to inform current policy choices, but also gives guidance as to how much of our resources we should devote to obtaining risk reduction. In this regard, the size of risk premium, aggregated over all members of society, measures the benefits of eliminating uncertainty, and therefore gives an upper bound on the value of resources that should be devoted to resolving the associated uncertainty. Again there is a subjective element here, since different agents will generally display different degrees of risk aversion.

34.2.7 Irreversibilities and Option Values

One fairly universal feature of uncertainty is that the farther we look into the future, the more uncertain we are about functional relationships and outcomes. As one consequence, uncertainty is a far bigger concern for dealing with policies involving the long-term future than for those that are atemporal in nature. Emerging research suggests that the resilience of nearly all environmental systems is characterized by linked dynamics across spatial and temporal scales, including processes that play out on very long time scales (Gunderson et al. 1995). Thus, it is vital to assess long-term consequences and deal with the attendant uncertainties. A critical factor in determining how we deal with the long-run future involves *reversibility*. Certain kinds of actions/policies are reversible in that if near-term consequences are not to our liking, we can take action that insulates the more distant future from these consequences. For example, in farming we may try a new fertilizer in one year. Unless this application has some long-term leaching effect on the soil the action is reversible in the next year.

The most difficult situations in this regard occur when there is at least some possibility that one could never return to the *status quo ante* no matter what reversing policies are used. Returning to the Gulf Stream circulation example, some have suggested that global warming could decrease salinity sufficiently in mid-Atlantic waters through arctic ice melt to shut off the pump entirely (Rahmstorf 1999). Should this happen, the event would be effectively irreversible and have very unpleasant long-term consequences for Northern Europe, which could, paradoxically, enter a cooling phase while the rest of the world warmed.

How general is the presence of potential irreversibility in our decision-making world? Although the relevant science is not generally well understood, several likely examples have emerged (Carpenter et al. 2001; Scheffer et al. 2001). For instance, phosphorous loading in a lake can cause it to flip to a "dirty" state in a way that cannot be reversed by simply eliminating the loading. Also, it is not yet clear whether or not fisheries collapse is reversible by simply stopping the fishing. Obviously species lost to extinction can never be recovered, though more needs to be known about whether there are commensurate irreversible losses of essential ecosystem services. Clearly it is of great importance that we improve our understanding of functional uncertainties in complex systems so as to avoid irreversible effects before they occur.

When irreversibilities are present, an argument can be made for some form of a "precautionary principle." In our example here, the potential of Gulf Stream 'shut-down' should make us especially cautious about overall global warming, which might trigger the sufficiently decreased salinity scenario. Note also that when there are potential irreversibilities, the rate of discount takes on special importance, since the weight put on long-run irreversible costs is very sensitive to the rate of discount. Consequently, we need to be particularly careful to defend our choice of discount rate and acknowledge the value judgments inherent in that choice.

Unfortunately, there is no single strategy for dealing with irreversibilities. The precautionary principle suffers from some of the same difficulties as the *avoidance* principle discussed earlier. Since we can never be sure what action or inaction will trigger an irreversible shift, we cannot be sure of what to be precautionary about. However, when learning more about the natural environment is an option, there are extra reasons to postpone actions whose desirability will depend on the information learned. In this case, the option to act differently depending on what is learned has value that is lost if we act irreversibly too soon. The potential for global climate change presents one situation where option values are important. Global climate change is characterized by unpredictable local or even regional extreme events, but we are as yet unable to foresee where and when these may occur. Evolution will be gradual, i.e. over decades to a century and is unpredictable. This gives us time for constructive learning, but only if we do not take irreversible actions (or inactions) now. (For a detailed discussion in the context of climate change see Heal and Kriström (2002): For a more basic review of environmental decision-making in the context of irreversibility, see Dasgupta and Heal (1979).

34.3 Conclusions and Recommendations

We conclude that advances and changes must be made in the way science is conducted and uncertainty communicated. Scientists must become more effective and compelling communicators of both what is and isn't known. Politicians must bolster their ability to make decisions in the face of uncertainty and be clear about the role ideology and values play in interpreting uncertainty. They must also be willing, at times, to risk votes in order to follow societally beneficial and scientifically assessed pathways. We must develop institutions that maintain flexibility, continually reassess, and potentially change direction in the presence of possible surprises and irreversibilities. And we must advance public understanding of the nature of complex problems with inherent uncertainties, and acceptance of the fact that solutions and policies must be ever changing as systems evolve and knowledge advances. To foster these broad aims, we offer specific recommendations (Box 34.2).

Box 34.2. Recommendations

Establish science-policy forums composed of leading figures in the science and policy communities. These forums should be constituted at the highest levels of government, within national academies of science, or within international science or policy bodies. These forums should provide input concerning the other recommendations outlined below and, further, provide oversight on the science-policy interface, offer constructive criticism on performance, and facilitate relevant communication links to the public at large.

Overhaul undergraduate and graduate education to include courses in complex systems, decision analysis, Bayesian statistics, and public communication.

Restructure scientific education at all levels to emphasize science as a process, and the Earth system as complex and, at times, highly uncertain in its dynamics.

Establish short courses, workshops, and institutes through which current scientists and policymakers could be brought up to speed on methods and knowledge relevant to their roles in the decision making process.

Educate natural-resource managers to view their systems as self-organizing, complex, and adaptive, with dynamics that play out on multiple scales of space and time, and across levels of organization. Train managers in the use of management 'experiments' that can be used to probe possible thresholds in the system, and distinguish between reducible and irreducible uncertainty.

Develop organizations and processes for dealing with environmental and natural-resource challenges that maintain flexibility, foster reassessment, and change direction if needed.

Science is largely a publicly supported enterprise, and we must ensure that it is serving the public. We simply must have professional scientists fully engaged in the policy process. They must recognize, however, that their expertise is in elucidating probabilities and consequences. When it comes to more value-laden discussions, their voices should carry no more weight than any other informed citizen. Thus scientists need to be fully aware of their own values and subjective assessments when engaged in this process. Not all scientists need engage in such activities, but all should support those reputably engaged in the policy process, and create the conditions that allow and value that engagement. To do otherwise defaults on the debt owed the public.

Politicians already have a well-developed ability to make decisions in the face of uncertainty – they do so everyday. They have been less successful, however, in elucidating the ideological filters through which uncertainty is interpreted. Risk assessment generally cannot be divorced from ideology; the risks we are willing to tolerate, or decide we must avoid at all costs, are matters of values. Politicians should be clear about the extent to which their decisions rest on a scientific assessment of probabilities and outcomes, and to what extent they rest on a more subjective and ideologically driven interpretation of how the world works. Policymakers owe this level of honesty to their constituents, who as they advance their own understanding of the complex workings of the world, should come to demand it.

The general public thus must also become much more aware of the nature of environmental challenges and the uncertainties plaguing them. It has, to some extent, been the ignorance of the public that has allowed politicians and other policymakers to invoke scientific uncertainties as a justification for inaction. Uncertainty cuts both ways – there is no certainty or safety in inaction. Society must cultivate an educated citizenry focused on both their rights and responsibilities. The public has an absolute right to be at the table in discussions of environmental challenges; they also have a responsibility to educate themselves on the issues before taking that seat. Scientists must also recognize the right that stakeholders have to be at the table, and the necessity of a broad perspective when it comes to negotiating the values we will hold and the futures we will pursue.

References

Berger, J. O. (1985). *Statistical decision theory and Bayesian analysis* (2nd ed.). New York: Springer.

Berkes, F., Colding, J., & Folke, C. (2002). *Navigating social-ecological systems*. Cambridge: Cambridge University Press.

Carpenter, S. R., Walker, B. H., Anderies, J. M., & Abel, N. (2001). From metaphor to measurement: Resilience of what to what? *Ecosystems, 4*, 765.

Dasgupta, P., & Heal, G. M. (1979). *Economic theory and exhaustible resources*. Cambridge: Cambridge University Press.

Funtowicz, S. O., & Ravetz, J. R. (1991). *Uncertainty and quality in science for policy, theory and decision library series A: Philosophy and methodology of the social sciences 15*. Dordrecht, the Netherlands: Kluwer Academic.

Gunderson, L. H., & Holling, C. S. (2002). *Panarchy: Understanding transformations in human and natural systems*. Washington, DC: Island Press.

Gunderson, L. H., Holling, C. S., & Light, S. S. (Eds.). (1995). *Barriers and bridges to the renewal of ecosystems and institutions*. New York: Columbia University Press.

Heal, G., & Kriström, B. (2002). Uncertainty and climate change. *Environmental Research and Economics, 22*, 3–39.

HELCOM. (2002). Environment of the Baltic Sea area 1994–1998. Baltic Sea Environment Proceedings No. 82B, p. 215.

Holmes, B. (1994). Biologists sort the lessons of fisheries collapse. *Science, 264*, 1252–1253.

Jackson, B. C., Kirby, M. X., Berger, W. H., Bjorndal, K. A., Botsford, L. W., Bourque, B. J., et al. (2001). Historical over-fishing and the recent collapse of coastal ecosystems. *Science, 293*, 629–660.

Lee, P. M. (1997). *Bayesian statistics: An introduction*. London: Arnold.

Mastrandrea, M., & Schneider, S. H. (2001). Integrated assessment of abrupt climatic changes. *Climate Policy, 1*(4), 433–49.

Morgan, M. G., & Henrion, M. (1990). Uncertainty: *A guide to dealing with uncertainty in Quantitative risk and policy analysis*. New York: Cambridge University Press.

Moss, R. H., & Schneider, S. H. (2000). Uncertainties in the IPCC TAR: Recommendations to lead authors for more consistent assessment and reporting. In R. Pachauri, T. Taniguchi, &

K. Tanaka, (Eds.). *Guidance Papers on the Cross Cutting Issues of the Third Assessment Report of the IPCC*, Intergovernmental Panel on Climate Change. Geneva, pp 31–51; Available from the Global Industrial and Social Progress Research Institute: http://www.gispri.or.jp

Pratt, J. W., Raiffa, H., & Schleiffer, R. (1995). *Introduction to statistical decision theory*. Cambridge, MA: MIT Press.

Press, S.J. (2003). Subjective and objective Bayesian statistics: Principles, models and applications, (2nd.ed.) Hoboken, NJ: Wiley-Interscience.

Rahmstorf, S. (1999). Shifting sea in the greenhouse? *Nature, 399*, 523–524.

Raiffa, H. (1968). *Decision analysis: Introductory lectures on choices under uncertainty*. Reading, MA: Addison-Wesley.

Scheffer, M., Carpenter, S., Foley, J., Folke, C., & Walker, B. (2001). Stochastic events can trigger large state shifts in ecosystems with reduced resilience. *Nature, 413*, 591–596.

Schneider, S. H. (2002). Editorial comment: Can we estimate the likelihood of climatic changes at 2100? *Climatic Change, 52*, 441–451.

Schneider, S. H., Turner, B. L., II, & Morehouse Garriga, H. (1998). Imaginable surprise in global change science. *Journal of Risk Research, 1*, 165–185.

Starrett, D. A. (1988). *Foundations of public economics*. New York: Cambridge University Press.

Chapter 35
Askö 2002: Commentary by Stephen Carpenter

Stephen Carpenter

Uncertainty must be considered by all forward-looking environmental assessments. Communication about uncertainty, among disciplines, policy makers, and the general public, turns out to be extraordinarily difficult. In this brief, clear article, Kinzig et al. analyze the difficulties and propose solutions. The article is a superb summary of current thinking about uncertainty. I routinely assign this article in my classes and find that students like it a lot. They know the topic is important, technical and complex. Yet the article is accessible, brief and clear. It is an excellent teaching tool.

What is uncertainty? In most cases, the space of future trajectories of social–ecological systems is unknown. See Fig. 35.1, which illustrates this unmeasured set of future possibilities. The set is unmeasured because the largest, most important changes in ecosystems have no historical precedent. The history of life on earth is one of emerging novelty and surprise, a one-way trip (Gould 1990). The same is true of the climate (Smith 2002) which is only one of many complex drivers of ecosystems and society. Nested within these unknowable changes is a region that could be known if we asked the right questions – the region of unasked questions (Carpenter et al. 2009). Within the region of unasked questions is the region of imaginable outcomes. This is the domain of scenarios, where provocative questions are asked even though the answers cannot be determined (Millennium Ecosystem Assessment 2005). Within the space of imaginable outcomes (accessible by scenarios) is the domain of recognized and quantifiable uncertainties. These recognized and quantifiable uncertainties surround a nugget of models and observations based on historical experience (Carpenter et al. 2006). To some extent, models and observations overlap and are mutually-reinforcing, but there are aspects of models that are not validated by data and observations that are not assimilated in models (Carpenter et al. 2006). The region of recognized uncertainties is the "known unknowns" that can be inferred from existing information. Within this region, it is possible to make predictions and uncertainty estimates based on historical experience.

S. Carpenter
Center for Limnology, University of Wisconsin, USA
e-mail: srcarpen@facstaff.wise.edu

T. Söderqvist et al. (eds.), *Bringing Ecologists and Economists Together*,
DOI 10.1007/978-90-481-9476-6_35, © Springer Science+Business Media B.V. 2011

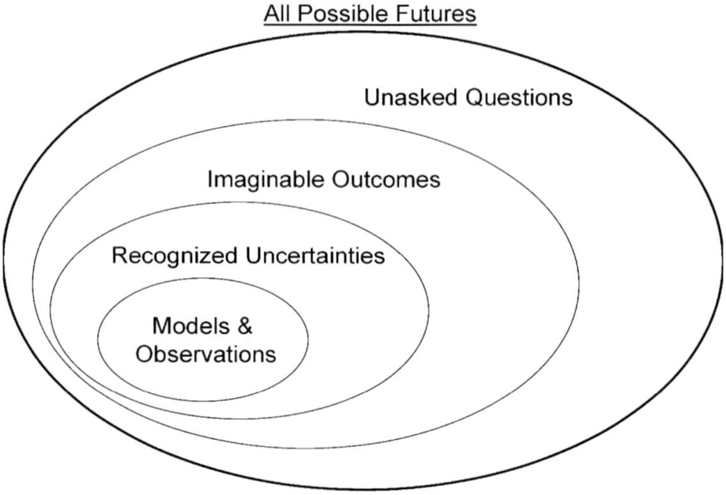

Fig. 35.1 The full set of possible futures of an ecosystem is only partially represented in available data and models. Together, the data and the models allow us to project the recognized uncertainties, or knowable unknowns. It is also possible to imagine outcomes that cannot be quantified (Carpenter et al. 2009). Beyond these are the unknowable unknowns (Smith 2002), the questions that have not been asked (Carpenter et al. 2009). The probability of any model projection of future ecosystem condition depends on the full set of possible futures, most of which is unknown. Adapted with permission from Carpenter et al. (2006)

Beyond the region are the "unknown unknowns" (the imaginable outcomes that cannot be quantified, the unasked questions, and the possible futures beyond the unasked questions), where probabilities cannot be computed because the space of potential outcomes is unmeasured.

Approaches to ecosystem management are different inside versus outside the region of recognized uncertainties (Carpenter 2002; Carpenter et al. 2006, 2009). Outside this region, it is misleading to think of management as the solution to a problem, but instead as the opportunity to search for potential solutions. These circumstances call for active research and experimentation. A few potential approaches include scenario analysis (Millennium Ecosystem Assessment 2005), adaptive ecosystem management (Gunderson and Holling 2002), resilience assessment (see Chapter 13 of this volume on Levin et al. 1998), formalizations of reasoning with imprecise probabilities (Walley 1991), and research to extend decision analysis to ambiguous (versus uncertain) situations (Epstein and Schneider 2005).

Kinzig et al. address situations within the region of recognized uncertainties, where we have access to historical experience and data, as well as rigorous and fairly complete theories of statistics, utility measurement and decision. In these circumstances, a fairly thorough assessment of uncertainty is possible. In spite of these advantages, there are significant difficulties in communicating the results to policymakers and the public, and in using the results in appropriate ways to evoke optimal social decisions (Sarewitz et al. 2000).

The 1997 floods in the Red River of North Dakota, U.S.A., are a classic example of miscommunication (Pielke 2003). Forecasters believed that the flood would exceed the previous record crest of 48.8 feet at Grand Forks, North Dakota. They predicted a crest of 49 feet, hoping to convey to decision makers that it would be the worst flood ever experienced in the region. Instead, local people were not alarmed. The 1979 flood had not been catastrophic, and the prediction for 1997 was only slightly higher, so why worry? On 22 April 1997, the Red River crested at 54 feet, inundating two cities and causing billions of dollars in damages. Many officials pointed to inaccurate forecasts as the cause of the disaster. However, historical analysis showed that flood forecasts were, on average, off by about 10%, so the 1997 forecast error was in line with historical experience. Instead, decision makers and forecasters failed to understand and communicate the uncertainty of the forecasts.

Such important breakdowns of communication are a matter of concern for Kinzig et al. Their analysis assumes that policy makers aim to maximize overall social utility which is measured in a fair, representative and inclusive way. This rational democratic ideal may rarely be achieved by political processes, but political distortions are not the focus of the paper. Instead, the authors are concerned with inadvertent distortions due to miscommunication among scientists and policy makers. In the rational democratic ideal, policy makers are most concerned about errors that harm people, their livelihoods, their security, or their environment. Policy makers therefore focus on the costs of particular errors, and are most concerned about avoiding the costly ones. Scientists, on the other hand, are most concerned about controlling or reducing average error. Kinzig et al. show how these different goals and attitudes toward error lead to miscommunication.

What can be done about it? Kinzig et al. propose forums of scientists and policy makers to develop guidelines for better communication of predictions and uncertainties. They propose educational reforms at every level to improve understanding of the environment, the nature of scientific uncertainty, and approaches for assessing the uncertainties that are important to people. If implemented, these recommendations would have far-reaching consequences. Nonetheless, scientists, managers and leaders have a further responsibility to push beyond the known unknowns, asking the unasked questions and expanding the space of imaginable outcomes.

References

Carpenter, S. R. (2002). Ecological futures: Building an ecology of the long now. *Ecology, 83*, 2069–2083.

Carpenter, S. R., Bennett, E. M., & Peterson, G. D. (2006). Scenarios for ecosystem services: An overview. *Ecology and Society, 11*(1), 29. [online] URL: http://www.ecologyandsociety.org/vol11/iss1/art29/.

Carpenter, S. R., Folke, C., Scheffer, M., & Westley, F. (2009). Resilience: Accounting for the noncomputable. *Ecology and Society, 14*(1), 13. [online] URL: http://www.ecologyandsociety.org/vol14/iss1/art13/.

Epstein, L. G., & Schneider, M. (2005). Learning under ambiguity. Manuscript available on the internet: http://www.stanford.edu/~schneidr/learn128.pdf.

Gould, S. J. (1990). *Wonderful life: The burgess shale and the nature of history* New York: Norton.

Gunderson, L. H., & Holling, C. S. (Eds.). (2002). *Panarchy*. Washington, DC: Island Press.

Levin, S., Barrett, S., Aniyar, S., Baumol, W., Bliss, C., Bolin, B., et al. (1998). Resilience in natural and socioeconomic systems. *Environment and Development Economics, 3*, 222–235.

Millennium Ecosystem Assessment. (2005). *Ecosystems and human well being: Scenarios*. Washington, DC: Island Press.

Pielke, R. A., Jr. (2003). The role of models in prediction for decision. In C. D. Canham, J. J. Cole, & W. K. Lauenroth (Eds.), *Models in ecosystem science* (pp. 111–138). Princeton, NJ: Princeton University Press.

Sarewitz, D., Pielke, R. A., Jr., & Byerly, R. (2000). *Prediction: science, decision making, and the future of nature*. Washington, DC: Island Press.

Smith, L. (2002). What might we learn from climate forecasts? *Proceedings of the National Academy of Sciences, 99*, 2487–2492.

Walley, M. P. (1991). *Statistical reasoning with imprecise probabilities*. London, UK: Chapmans Hall.

Chapter 36
Askö 2002: Commentary by Claude Henry

Claude Henry

Ann Kinzig, David Starrett and their coauthors forcefully call for a broad approach to uncertainty. In this context it might be useful to examine what may specifically be done, in both conceptual and operational terms, when uncertainty takes forms that are not easily formalized. In his *Treatise on Probability*, published in 1921, John Maynard Keynes makes a clear distinction between two kinds of uncertainty: the first one may be characterized by probabilities[1], while this is not possible for the second one. A similar distinction is made by Frank Knight in *Risk, Uncertainty and Profit*, published the same year. In what follows, when dealing with economic issues, we will use the knightian terms risk and uncertainty, and concentrate on uncertainty. However we will first contrast two cases drawn from the natural sciences.

36.1 Probabilistic vs. Uncertain Science

In large colliders like the huge ring at the CERN in Geneva, beams of particles (neutrons for example) are accelerated to very high speeds, hence very high levels of energy, and then hit a target made of heavy atoms. The atoms are disintegrated and the fragments are thrown out in various directions at various speeds. Before the collisions take place, it is possible to predict neither where the fragments produced by one well identified atom will go nor at which speeds they will travel. However quantum mechanics predicts with perfect accuracy the probabilities attached to their directions and speeds; these probabilities materialize in the collider as statistical frequencies of realized events.[2] Uncertainty in quantum mechanics is strictly

[1] Keynes doesn't introduce the distinction between objective and subjective probabilities.

[2] In that sense, they are objective probabilities.

C. Henry
School of International and Public Affairs, Columbia University, USA,
and Laboratoire d'Econométrie, Ecole Polytechnique, France
e-mail: ch2396@columbia.edu

T. Söderqvist et al. (eds.), *Bringing Ecologists and Economists Together*,
DOI 10.1007/978-90-481-9476-6_36, © Springer Science+Business Media B.V. 2011

probabilistic, and Werner Heisenberg, had he been an economist, would have called "principle of risk" his famous Uncertainty Principle.

Let us now consider a completely different situation in biology. In 1986, the bovine spongiform encephalopathy (BSE) was registered for the first time in English cows. Then the disease spread at an astonishing pace. But according to the government at the time, "British beef is safe", no contagion to humans is possible thanks to the "barrier between species". This pseudo-scientific concept, more or less relevant for other ailments, had no theoretical foundation and had of course not been experimentally tested in the case of BSE. The test came in 1991, when scientists at the University of Bristol succeeded in inoculating a cat with BSE; as is well known, just one counterexample is enough to destroy a hypothesis, here the existence of a barrier between species against BSE. The British government and its experts were left without any science, not even with uncertain science, not even an illusion of science, at a moment when they had to face a more sinister threat, a new form of the Creutzfeldt–Jakob disease (CJD) striking young people (the original form appears at older ages).

Both BSE and CJD are degenerative diseases of the brain. Could they be linked? A doctor in medicine and biochemist, Stanley Prusiner, happened to be working on that question at the University of California. More exactly he was working on neurodegenerative diseases that seemed to result from disorders of protein conformation: it seemed that proteins of the prion type could, after some mutations, become infectious pathogens of the brain. Dr. Prusiner had significant experimental results on mice and had elucidated some of the molecular events responsible for the mutations of the prions and for their infectious power.[3] The foundations were by no means complete – the experimental data were still limited and not all the relevant molecular events had been elucidated – but they appeared to constitute a significant subset of an ideally satisfactory set of experimental data and theoretical elements; and they were coherent. What Dr. Prusiner proposed at the time[4] was a piece of nonprobabilistic uncertain science, however sufficiently convincing to scientifically support the decision to bar English[5] beef from being consumed in the European Union.

36.2 The Concept of Reliable Uncertain Science

Economic, political and social decisions must often be made despite the fact that the available relevant information is uncertain. To decide under uncertain scientific information is not exceptional.[6] But not every piece of uncertain science is sufficiently well founded, theoretically and experimentally, to be credible, to be worth considering in a decision process; that was the case with the "barrier between species" concept in the BSE context. What uncertain science is to be considered and

[3] Never before had one observed proteins turned into infectious agents.

[4] He subsequently made progress to more complete theoretical foundations vindicated by relevant experimental data. For this work, he got the Nobel Prize in Medicine as early as 1997.

[5] BSE never spread to Scotland.

[6] As doctors in medicine well know.

mobilized for what kind of decision? We will explore this question in the framework of the famous, and infamous, asbestos saga.[7]

In 1898, about 20 years after asbestos started to be mined and processed, health inspectors, monitoring working conditions in various British industrial environments, observed strong correlations between the prevalence of certain lung diseases (that they were unable to characterize) and the presence of asbestos dust in workshops. In 1906, French health inspectors got similar results.

In 1911, British scientists exposed rats to asbestos dust: the rate of mortality was strongly linked to the density of dust, and respiratory troubles were involved.

This body of scientific information was thus already significant, but obviously rather incomplete. To insurance companies, they appeared sufficiently credible to warrant a ban on covering workers exposed to significant asbestos dust densities. However, public regulators, lobbied by industry, accepted that densities of asbestos in working environments had sufficiently decreased since the inception of the industry to be considered as harmless.

Twenty to 25 years later, this proved wrong.[8] And at that time, scientists were able to characterize the diseases, from which the workers exposed to asbestos were dying, as specific forms of cancer of the pleura. True, not every step in the development of such cancers, not all causal links, had been elucidated, but it is difficult to deny that the scientific knowledge available was more than adequate to justify a ban on asbestos itself. That had to wait for another 50 years, with human and economic consequences that are well documented.

In both cases, BSE/CJD and asbestos, a piece of uncertain science not only becomes less uncertain, but even tends to certainty, as time passes and more research is done. But it was not optimal to wait for scientific certainty before making crucial decisions. It other contexts, scientific certainty should not even be expected at any time; scientific information may improve while remaining irreducibly uncertain; then waiting for certainty not only is non-optimal but even meaningless, and the concept of credible, however uncertain, science is even more critical.

Consider for example how Henry Pollack, a professor of Earth Sciences at the University of Michigan, sees uncertainty stemming from complexity and considers as unavoidable the recourse to models in ecology, as in climatology, oceanography, etc.: "Because of the complexity, it is extremely difficult for even the most capable ecologists to study a forest ecosystem in its full detail, and so they develop simplified concepts about the workings of the ecosystem, focusing on a few components and their interactions that are thought to be particularly significant. This conceptualization of the ecosystem web of interactions is called a model. To be sure, different ecologists may perceive the interactions differently, weigh the participation of the different components differently and, therefore, develop different models. Because of the complexity, the ecosystem is imperfectly understood and uncertainty about how it all hangs together is attendant" (Pollack 2003, p. 106).

---

[7] See European Environment Agency (2001) for a more detailed chronology.

[8] In the meantime, insurance companies, influenced by the position public regulators had taken, lifted their ban. Ultimately, they would have to regret this move.

Pollack's illustration of system complexity and of the ensuing intrinsic limits on knowledge may easily be transposed to the scientific approach to climate change. Here the IPCC[9], a large network of scientists working under the aegis of the United Nations and the World Meteorological Organization, collect available relevant data from all over the world, encourage the production of useful missing data, and mobilize their members (experts in physics, chemistry, biology, ecology, economics, etc.) to evaluate the data, build models on that basis, and in this way characterize to the best of the available (uncertain) knowledge the perspectives of climate change. The process is continuous, but the IPCC official reports are published every 5–6 years.

In preparing the report published in 2001, the IPCC have distinguished six greenhouse gas emissions scenarios in the context of several climatic models. Their methodology led to conclusions such as "between 1990 and 2100, the average global temperature on earth will increase by 1.4°C to 5.8°C".[10] This range is the consequence of the diversity in the underlying scenarios and models. No probability distribution can meaningfully be associated to the interval [1.4, 5.8]. In this sense, the science of climate change is genuinely uncertain. Is it credible?

It seems clear that the IPCC's process of investigation contributes to the theoretical and empirical scientific knowledge of climate change in a rigorous, systematically organized and well-monitored way. The physicist and historian of science John Ziman characterizes scientific credibility in the following way: "The credibility of science depends as much on how it operates as a collective social enterprise, as it does on the principles regulating the type of information that this enterprise accepts and transforms into knowledge" (Ziman 2000, p. 13). It seems that the work of the IPCC is favorably placed with respect to both these criteria, and that their findings constitute a credible basis justifying a significant reduction in greenhouse gas emissions.

However, John Ziman makes a warning, reminiscent of Thomas Kuhn's *The Structure of Scientific Revolutions*: "Normal science is a mindset that can take hold of researchers in almost any field of academic science". And he explains: "It is the rationale of laboratory life, ensuring that order emerges from region of disorder and that communications from laboratories with different signatures can be combined coherently" (Ziman 2000, p. 201). Would it be the case that normality of science has made the IPCC models blind to such developments as the unexpectedly rapid meltdown of the Arctic ice sheet,[11] or the life-threatening acidification of oceans, or some other feed-back effects of CO_2 emissions? This adds to uncertainty, but does certainly not diminish the need for a significant reduction in greenhouse gas emissions.

However, this scientific basis obviously does not possess all the attributes of canonical (deterministic or probabilistic) science. It was on this ground that the US

[9] Intergovernmental Panel on Climate Change.

[10] It would be significantly more locally, two to three times more in Arctic regions, for example.

[11] According to results gathered in a report from the NASA's Goddard Institute for Space Studies.

government rejected the conclusions of the IPCC as scientifically unsound.[12] In 2003 the Bush administration produced its own review of the perspectives of climate change. This review was then examined by a panel of experts put together by the National Academy of Sciences. According to *The New York Times* (March 1, 2003), "The panel described Mr. Bush's plan as a redundant examination of issues that had largely been settled; bereft of vision, executable goals and timetables; in short, little more than a cover-up for inaction".

36.3 From Decision Theory Under Risk to Decision Theory Under Uncertainty

As we have seen, it appears possible to frame decision-making under uncertain information, but not with the degree of accuracy made possible with the instruments provided by John von Neumann, Oskar Morgenstern and Leonard Savage[13] when uncertainty boils down to risk. The range of applications that stem from von Neumann–Morgenstern's contribution is stunning: in finance, in insurance, in investments' appraisal, in resource and environmental economics, etc. Within economics, and more largely within social sciences, it has been a "scientific revolution".[14] However, any scientific revolution, however deep and influential, some day reaches its limits. Here, limits were revealed by Maurice Allais and David Ellsberg[15] under the form of "paradoxes", i.e., contradictions between predictions within the von Neumann–Morgenstern model (extended by Savage) and observed non-pathological behaviors with respect to risk. Ellsberg's experiments[16] were particularly illuminating, insofar as they elicited standard behaviors that were incompatible with the use (explicit or implicit) of probabilities by the persons submitted to the experiences.

Allais's and Ellsberg's paradoxes convinced some mathematicians and economists, especially Itzhak Gilboa, Mark Machina and David Schmeidler, to find ways of weakening some assumptions supporting the von Neumann–Morgenstern model and of still having a manageable model that formalizes a large range of behaviors, including those considered in Allais's and Ellsberg's paradoxes. Other pioneers are Kenneth Arrow and Leonid Hurwicz. In Arrow and Hurwicz (1977), they build on Shackle (1949) to deal with decision contexts such as the following one: "There are two coins, but coin two is flipped only if coin one shows tails" (Arrow and Hurwicz

[12] For example, in a speech at the American Academy in Berlin (September 17, 2003), Energy Secretary Spencer Abraham required "sound science" to "justify further action" in the matter of climate change.

[13] See von Neumann and Morgenstern (1944) and Savage (1954). See also Kinzig et al. in Chapter 34.

[14] In the sense of Kuhn (1962).

[15] Maurice Allais is a well-known economist, who was Nobel Laureate in 1988. David Ellsberg's contributions to economics are more limited, but the man got famous in the USA when he leaked the *Pentagon Papers*, during the Vietnam war.

[16] Reported in Ellsberg (1963).

1977, p. 463). This is indeed a context that one encounters in the most recent papers about decisions under genuine uncertainty. The most general results in this spirit can be found in Maccheroni et al. (2006) and Klibanoff et al. (2005); the latter opens the way to econometric applications.

These authors basically show the existence, for a person confronted to decisions under uncertainty, of a convex set P of probability distributions on the set X of possible outcomes, and of a real-valued function u on X (analogous to the function that plays a central role in the von Neumann–Morgenstern model). P and u are such that the preferences of the concerned person on the random outcomes $\tilde{x}$ (i.e., the uncertain outcomes taking a random value in X) can be represented by a weighted sum of mathematical expectations in $\tilde{x}$:

$$\alpha \min_{\mu \in P} \mathrm{E}_{\mu}\left[u(\tilde{x})\right] + (1-\alpha) \max_{\mu \in P} \mathrm{E}_{\mu}\left[u(\tilde{x})\right].$$

In this stochastic utility function, α is the *ambiguity aversion index*; it lays in the interval [0,1] and its value in this interval is endogenous, in the sense that it is an output of the theorem; α numerically characterizes the person's aversion for uncertainty, which runs deeper than risk aversion.

What is the link between this formalization and the decision processes pertaining to the BSE in England, to asbestos in Europe and the USA, and to the prospects of climate change worldwide? In each of these cases, one may figure a set of events, each event being characterized by a certain content of more or less uncertain scientific information that is suitably summarized by a non-additive set function. There exists a subclass Γ (possibly empty) on which the set function is indeed additive. This subclass may be called the subclass of unambiguous events as scientific information there rests on a unique probability distribution. Acts which operate only on unambiguous events (i.e., are measurable with respect to Γ) may be termed unambiguous acts.

If a decision-maker a priori rejects as "scientifically unsound" any act which is not unambiguous, that means that he sticks to the maximization of a von Neumann–Morgenstern expected utility on the set of acts which are scientifically unambiguous. In so doing, he neglects a large array of scientific information which, however uncertain, might be reliable and decisive; that happened in the asbestos case, but not in the BSE one (to the credit, rarely recognized, of the UK Government); it seems on its way to happen as regards climate change.

In short, we can say that "optimizing on the set of acts which are scientifically unambiguous is not optimal". This is the least controversial statement one can give of the famous *precautionary principle* (see Henry and Henry 2002). In other words, precaution requires that the decision-maker optimizes on a set larger than the set of scientifically unambiguous acts, as should have been done in the 1940s–1950s for asbestos, and has rightly been done for BSE in 1993. When the relevant conditions are satisfied, the optimization may be performed using as maximand the weighted expected utility derived by Klibanoff, Maccheroni and their co-authors.

It is obvious that this model of decision-making under uncertainty is more complicated to use in applications than the standard von Neumann–Morgenstern one. However, under more or less sophisticated forms, it has already been used to explain some otherwise puzzling instances of incompleteness in financial markets (Mukerji and Tallon 2001), to deal with imprecise probabilistic information (Gajdos et al. 2004), to overcome obstacles against the introduction of new financial instruments meant at insuring against catastrophic events (Bantwal and Kunreuther 2000), to introduce a measure of coherence in the approaches to the *precautionary principle* as we have seen above (Henry and Henry 2002), etc.

36.4 Conclusion

We might conclude tentatively that the Keynes–Knight distinction between risk and uncertainty is fruitful under various methodological approaches and in a broad range of applications. However, uncertainty should not be inflated and invoqued as an alibi for inaction; in 2003 the National Academy of Sciences brought home the point to President Bush. Inaction, despite increasingly reliable uncertain scientific information, bordering to certainty from the 1950s on, had and has catastrophic consequences for tens of thousands of victims of asbestos. And it might soon be the same story for billions of victims of climate change.

Climate science is indeed uncertain, but less so as time passes and as research brings more compelling results. Its might even be the case that we already have enough quasi-certainties at hand to warrant far reaching changes in our economic, political and social behaviors and institutions. This is, among others, the conclusion that Elizabeth Kolbert, journalist at *The New Yorker*, draws from a broad inquiry, involving politicians in Washington, scientists at NASA, Princeton, Harvard and other universities and research centers, Alaskans, Inuits and Siberians on the front of the changes, etc.: "As the effects of global warming become more and more difficult to ignore, will we react by finally fashioning a global response? Or will we retreat into ever narrower and more destructive forms of self-interest? It may seem impossible to imagine that a technologically advanced society could chose, in essence, to destroy itself, but that is what we are now in the process of doing" (Kolbert 2006). The French economist and philosopher Jean-Pierre Dupuy, professor at Ecole Polytechnique in Paris and at Stanford University, goes even deeper: "To turn around what we identify with progress, would have such phenomenal consequences that we refuse to believe what we indeed know as being true" (Dupuy 2002, p. 144).

Acknowledgements I am grateful to Kenneth Arrow and Marc Henry for valuable comments and suggestions on a previous draft. I retain sole responsibility for the version presented here.

References

Arrow, K. J., & Hurwicz, L. (1977). An optimality criterion for decision-making under ignorance. In K. J. Arrow & L. Hurwicz (Eds.), *Studies in resource allocation processes* (pp. 463–471). Cambridge, UK: Cambridge University Press.

Bantwal, V., & Kunreuther, H. (2000). A cat bond premium puzzle? *The Journal of Psychology and Financial Markets, 1*, 76–91.

Dupuy, J.-P. (2002). *Pour un catastrophisme éclair – Quand l'impossible est certain*. Paris: Editions du Seuil.

Ellsberg, D. (1963). Risk, ambiguity and the savage axioms. *Quarterly Journal of Economics, 75*, 643–669.

European Environment Agency. (2001). Late lessons from early warnings: The precautionary principle 1896–2000. Environmental Issue Report no. 22, EEA, Copenhagen.

Gajdos, T., Tallon, J.-M., & Vergnaud, J.-C. (2004). Decision making with imprecise probabilistic information. *Journal of Mathematical Economics, 40*, 647–681.

Henry, C., & Henry, M. (2002). Formalization and applications of the precautionary principle. Cahier du Laboratoire d'Econométrie de l'Ecole Polytechnique, no. 2002–008.

Keynes, J. M. (1921). *A treatise on probability*. London: Macmillan.

Klibanoff, P., Marinacci, M., & Mukerji, S. (2005). A smooth model of decision making under ambiguity. *Econometrica*, 73, 1849–1892.

Knight, F. (1921). *Risk, uncertainty and profit*. New York: Kelly.

Kolbert, E. (2006). *Field notes from a catastrophe – Man, nature and climate change*. New York: Bloomsbury.

Kuhn, T. S. (1962). *The structure of scientific revolutions*. Chicago: The University of Chicago Press.

Maccheroni, F., Marinacci, M., & Rustichini, A. (2006). Ambiguity aversion, robustness and the variational representation of preferences. *Econometrica, 74*, 1447–1498.

Mukerji, S., & Tallon, J.-M. (2001). Ambiguity aversion and incompleteness of financial markets. *Review of Economic Studies, 68*, 883–904.

Pollack, H. (2003). *Uncertain science...uncertain world*. Cambridge: Cambridge University Press.

Savage, L. (1954). *The foundations of statistics*. New York: Wiley.

Shackle, G. L. S. (1949). *Expectation in economics*. Cambridge, UK: Cambridge University Press.

von Neumann, J., & Morgenstern, O. (1944). *Theory of games and economic behavior*. Princeton: Princeton University Press.

Ziman, J. (2000). *Real science – What it is, and what it means*. Cambridge: Cambridge University Press.

Part IV
Conclusion

Chapter 37
The Askö Challenge

Carl Folke

Back in 1982, AnnMari Jansson, Bengt-Owe Jansson and Karl-Göran Mäler gathered leading economists and systems ecologists in Saltsjöbaden outside Stockholm, to discuss barriers and bridges between the two disciplines and ways to move forward toward improved stewardship of natural resources, ecosystems and the environment (Jansson 1984). Major differences in perspectives and approaches caused tensions and arguments and a gulf of disagreements. This first encounter was the seed for the establishment of the International Society for Ecological Economics in 1989. The hope was to create a scientific forum for interdisciplinary dialogue and collaboration on sustainability issues between natural and social scientists, predominantly ecologists and economists (e.g., Costanza 1989). AnnMari Jansson played a key role as one of the founders, together with Bob Costanza, Herman Daly and David Pearce and I became part of this new effort. The concept of Natural Capital was essential in this context, consisting of non-renewable resources, renewable resources and ecosystem services, the latter two generated by the life-supporting ecosystems (Odum 1989; Jansson et al. 1994).

In parallel two subfields of economics – natural resource economics and environmental economics – were coming together (e.g., Fisher 1981; Perrings 1987; Barbier 1989), reflected for example in the establishment in 1990 of the European Association of Environmental and Resource Economists, and the recognition that resources and the environment are two sides of the same coin and tightly coupled to development and human welfare (Pearce and Turner 1990; Dasgupta and Mäler 1991).

It was during this period, in 1991, that the Royal Swedish Academy of Sciences decided to launch the Beijer International Institute of Ecological Economics, with the mission to carry out research and stimulate scientific collaboration to promote a deeper understanding of the interplay between ecological systems and social and

C. Folke
Beijer Institute of Ecological Economics, the Royal Swedish Academy of Sciences, Box 50005, SE-104 05 Stockholm, Sweden and Stockholm Resilience Centre, Stockholm University, SE-106 91 Stockholm, Sweden
e-mail: carl.folke@beijer.kva.se

T. Söderqvist et al. (eds.), *Bringing Ecologists and Economists Together*,
DOI 10.1007/978-90-481-9476-6_37, © Springer Science+Business Media B.V. 2011

economic development. Karl-Göran Mäler became the first Director and asked me to join as his Deputy. Was it really possible to combine ecology and economics? Different mental models, perspectives, opinions – how could we find a way to make skilful economists and ecologist start talking to each other? We needed to create a platform for constructive dialogue and trust building, an open space where we could actively search for a deeper understanding of the foundation for a prosperous societal development, for sustainable futures. The Askö meeting was invented, an annual meeting taking place at the marine field station Askö in the Stockholm archipelago, a station dear to Bengt-Owe and AnnMari Jansson.

The Askö meetings cover a diverse set of issues as reflected in this volume, some with more of an economic focus and others more ecological and some truly a reflection of integrative science, but all with insights from both areas. The ten contributions reprinted here reflect this diversity published in *Ambio*, *BioScience*, *Environment and Development Economics*, *Environmental Science and Technology*, *Journal of Economic Perspectives*, *Resource and Energy Economics* and *Science*. Not surprisingly, the papers in Science are those that have got the most attention on the citation market (Table 37.1) and two of those are continuously used in research and discussion about economic development, life-support systems and Kuznets curves (Arrow et al. 1995) and in the explosion of research and policy on the significance and values of ecosystem services (Daily et al. 2000). We have continued the Askö tradition and every year economists and ecologists gather at the beautiful island talking about complex adaptive systems, ecosystem services and their governance, the ethics of the environment, uncertainty and surprise and looming global challenges.

These face-to-face meetings of ecologists and economists, and occasionally also other natural and social scientists, broadened the within-disciplinary perspectives of many and clarified potentials and limitations, not necessarily always leading to consensus on the issues, methods and assumptions. But, more importantly people gained a broader systems view of their own research and started to rephrase it in a more general and, as many would argue, more realistic context. Research moved from incorporating insights from other disciplines into one's own research field to a deeper understanding of the combined system of humans and nature. It became clear that humans and nature are not two separate parts, but truly integrated and interdependent;

Table 37.1 Citations in Web of Science/ISI, Google Scholar and Scopus of published Askö papers reprinted in this book, mid-May 2009

Askö paper	Journal	ISI	Google Scholar	Scopus
Arrow et al. (1995)	*Science*	345	875	375
Levin et al. (1998)	*Environment and Development Economics*	30	90	–
Daily et al. (1998)	*Science*	35	56	44
Daily et al. (2000)	*Science*	110	210	125
Arrow et al. (2000)	*Environmental Science and Technology*	23	39	23
Kinzig et al. (2003)	*Ambio*	35	45	35
Arrow et al. (2004)	*Journal of Economic Perspectives*	50	145	55
Heal et al. (2004)	*Resource and Energy Economics*	5	15	5

that the biosphere and its life-support systems provide the biophysical foundation for human wellbeing; that preferences, knowledge systems, incentives, institutions and governance systems, and their underlying world views, shape the capacity of ecosystems to sustain societal development; that choices, decisions and policies directly and indirectly affect the trajectories and pathways of social–ecological systems and that some of those pathways may be irreversible.

There were tensions, lively discussion and debates but no real fights or major conflicts. Dynamic discussions and differences of opinions created space for new thoughts, new inquiries and strands of research and increased the humbleness of those involved – it was sometimes quite exhausting but always immensely rewarding. We were lucky to have wonderful people involved!

Research institutes like the Beijer play a significant role as catalysts for research activities not easily initiated within university departments. They have the potential to speed up the process of learning for sustainability. I am privileged for having the opportunity to work at such a dynamic institute, interacting with great people and great scholars, under the support of the Royal Swedish Academy of Sciences and the Beijer Foundation and now engaging Beijer's legacy and work into the Stockholm Resilience Centre. The Askö experience will continue as a crucial part of this journey!

References

Arrow, K., Bolin, B., Costanza, R., Dasgupta, P., Folke, C., Holling, C. S., et al. (1995). Economic growth, carrying capacity, and the environment. *Science, 268*, 520–521.

Arrow, K., Daily, G. C., Dasgupta, P., Levin, S., Mäler, K.-G., Maskin, E., et al. (2000). Managing ecosystem resources. *Environmental Science and Technology, 34*, 1401–1406.

Arrow, K., Dasgupta, P., Goulder, L., Daily, G. C., Ehrlich, P., Heal, G., et al. (2004). Are we consuming too much? *Journal of Economic Perspectives, 18*, 147–172.

Barbier, E. B. (1989). *Economics, natural resource scarcity and development: Conventional and alternative views*. London: Earthscan.

Costanza, R. (1989). What is ecological economics? *Ecological Economics, 1*, 1–7.

Daily, G., Dasgupta, P., Bolin, B., Crosson, P., Du Guerny, J., Ehrlich, P., et al. (1998). Food production, population growth, and the environment. *Science, 281*, 1291–1292.

Daily, G., Söderqvist, T., Aniyar, S., Arrow, K., Dasgupta, P., Ehrlich, P. R., et al. (2000). The value of nature and the nature of value. *Science, 289*, 395–396.

Dasgupta, P., & Mäler, K.-G. (1991). The environment and emerging development issues. In *Proceedings of the World Bank annual conference on development economics 1990* (pp. 101–131). The International Bank for Reconstruction and Development/The World Bank.

Fisher, A. C. (1981). *Resource and environmental economics*. Cambridge, UK: Cambridge University Press.

Heal, G., Walker, B., Levin, S., Arrow, K., Dasgupta, P., Daily, G. C., et al. (2004). Genetic diversity and interdependent crop choices in agriculture. *Resource and Energy Economics, 26*, 175–184.

Jansson, A. M. (Ed.). (1984). Integration of economy and ecology: An outlook for the eighties. *Proceedings from the Wallenberg Symposia* (pp. 19–29). Stockholm: Askö Laboratory, University of Stockholm.

Jansson, A. M., Hammer, M., Folke, C., & Costanza, R. (Eds.). (1994). *Investing in natural capital: The ecological economics approach to sustainability*. Washington, DC: Island Press.

Kinzig, A., Starrett, D., Arrow, K., Bolin, B., Dasgupta, P., Ehrlich, P., et al. (2003). Coping with uncertainty: A call for a new science-policy forum. *Ambio, 32*, 330–335.

Levin, S. A., Barrett, S., Aniyar, S., Baumol, W., Bliss, C., Bolin, B., et al. (1998). Resilience in natural and socioeconomic systems. *Environment and Development Economics, 3*, 222–235.

Odum, E. P. (1989). *Ecology and our endangered life-support systems*. Sunderland, MA: Sinuaer Associates.

Pearce, D. W., & Turner, R. K. (1990). *Economics of natural resources and the environment*. London: Harvester-Wheatsheaf.

Perrings, C. (1987). *Economy and environment*. Cambridge: Cambridge University Press.

Appendix: List of Askö Meetings, Participants and Publications 1993–2002

The 1st Askö Meeting, 1993:

Robert Costanza
Partha Dasgupta
Paul Ehrlich
Carl Folke
Geoffrey Heal
Buzz Holling
AnnMari Jansson
Bengt-Owe Jansson
Bengt Kriström
Jonas Larsson
Karl-Göran Mäler
Charles Perrings
Jonathan Roughgarden
David Starrett
Julie Sweitzer

Askö 1993. *Beijer Reprint No. 46*: Folke, C. (1995). Ecologists and economists can find common ground. *Bioscience, 45*, 283–284.

The 2nd Askö Meeting (at Nässlingen), 1994:

Kenneth Arrow
Bert Bolin
Robert Costanza
Partha Dasgupta
Carl Folke
Buzz Holling
Bengt-Owe Jansson
Simon Levin

DOI 10.1007/978-90-481-9476-6, © Springer Science+Business Media B.V. 2011

Karl-Göran Mäler
Charles Perrings
David Pimentel

Askö 1994. *Beijer Reprint No. 48:* Arrow, K., Bolin, B., Costanza, R., Dasgupta, P., Folke, C., Holling, C. S., Jansson, B.-O., Levin, S., Mäler, K.-G., Perrings, C., & Pimentel, D. (1995). Economic growth, carrying capacity, and the environment. *Science, 268*, 520–521.

The 3rd Askö Meeting, 1995: The Resilience of Social and Biological Systems

Sara Aniyar
Scott Barrett
William Baumol
Christopher Bliss
Bert Bolin
Partha Dasgupta
Paul Ehrlich
Carl Folke
Ing-Marie Gren
Buzz Holling
AnnMarie Jansson
Bengt-Owe Jansson
Simon Levin
Dan Martin
Karl-Göran Mäler
Charles Perrings
Eytan Sheshinski

Askö 1995. *Beijer Reprint No. 108:* Levin, S., Barrett, S., Aniyar, S., Baumol, W., Bliss, C., Bolin, B., Dasgupta, P., Ehrlich, P., Folke, C., Gren, I.-M., Holling, C. S., Jansson, A., Jansson, B.-O., Mäler, K.-G., Martin, D., Perrings, C., & Sheshinski, E. (1998). Resilience in natural and socioeconomic systems. *Environment and Development Economics, 3*, 222–235.

The 4th Askö Meeting, 1996: The Environment and Ethics

Kenneth Arrow
Bert Bolin

Johan Colding
Partha Dasgupta
Paul Ehrlich
Carl Folke
Clifford Geerts
Ing-Marie Gren
James Griffin
Buzz Holling
Carl-Olof Jacobson
AnnMari Jansson
Bengt-Owe Jansson
Simon Levin
Assar Lindbeck
Karl-Göran Mäler

The 5th Askö Meeting, 1997: Population, Food Availability and the Environmental Resource Base

Bert Bolin
Pierre Crosson
Gretchen Daily
Partha Dasgupta
Jacques du Guerny
Paul Ehrlich
Carl Folke
AnnMari Jansson
Bengt-Owe Jansson
Nils Kautsky
Ann Kinzig
Simon Levin
Karl-Göran Mäler
Per Pinstrup-Andersen
Domenico Siniscalco
Brian Walker

Askö 1997. *Beijer Reprint No. 103:* Daily, G. C., Dasgupta, P., Bolin, B., Crosson, P., du Guerny, J., Ehrlich, P., Folke, C., Jansson, A., Jansson, B.-O., Kautsky, N., Kinzig, A., Levin, S., Mäler, K.-G., Pinstrup-Andersen, P., Siniscalco, D., & Walker, B. (1998). Food production, population growth, and the environment. *Science, 281*, 1291–1292.

The 6th Askö Meeting, 1998: Valuing Ecosystem Services

Sara Aniyar
Kenneth Arrow
Gretchen Daily
Partha Dasgupta
Paul Ehrlich
Carl Folke
AnnMari Jansson
Bengt-Owe Jansson
Nils Kautsky
Simon Levin
Jane Lubchenco
Karl-Göran Mäler
David Simpson
Tore Söderqvist
David Starrett
David Tilman
Brian Walker

Askö 1998. *Beijer Reprint No. 132:* Daily, G. C., Söderqvist, T., Aniyar, S., Arrow, K., Dasgupta, P., Ehrlich, P., Folke, C., Jansson, A., Jansson, B.-O., Kautsky, N., Levin, S., Lubchenco, J., Mäler, K.-G., Simpson, D., Starrett, D., Tilman, D., & Walker, B. (2000). The value of nature and the nature of value. *Science, 289*, 395–396.

Off-site Askö Meeting, Washington, 1999: Institutions for Managing Ecosystem Services

Kenneth Arrow
Gretchen Daily
Partha Dasgupta
Carl Folke
Simon Levin
Eric Maskin
Karl-Göran Mäler
David Starrett
Thomas Sterner
Tom Tietenberg

Askö in Washington 1999. *Beijer Reprint No. 133:* Arrow, K., Daily, G. C., Dasgupta, P., Levin, S., Mäler, K.-G., Maskin, E., Starrett, D., Sterner, T., & Tietenberg, T. (2000). Managing ecosystem resources. *Environmental Science and Technology, 34*, 1401–1406.

The 7th Askö Meeting, 1999: Evolution in Biological, Social and Economic Systems

Sara Aniyar
Paul Ehrlich
Peter Hammerstein
AnnMari Jansson
Bengt-Owe Jansson
Nils Kautsky
Simon Levin
Jane Lubchenco
Karl-Göran Mäler
Eric Maskin
Erling Norrby
Paul Portney
Leslie Real
Mari Rege
Domenico Siniscalco
David Starrett
Stephen Stearns
Brian Walker
Jörgen Weibull

Off-site Askö Meeting, Stanford, 2000: Consumption

Kenneth Arrow
Gretchen Daily
Partha Dasgupta
Anne Ehrlich
Paul Ehrlich
Lawrence Goulder
Geoff Heal
Simon Levin
Karl-Göran Mäler
Stephen Schneider
David Starrett
Brian Walker

Askö in Stanford 2000. *Beijer Reprint No. 209:* Arrow, K., Dasgupta, P., Goulder, L., Daily, G. C., Ehrlich, P., Heal, G., Levin, S., Mäler, K-G., Schneider, S., Starrett, D., & Walker, B. (2004). Are we consuming too much? *Journal of Economic Perspectives, 18*, 147–172.

The 8th Askö Meeting, 2000: Environment, Development and Trade

Sara Aniyar
Kenneth Arrow
Bert Bolin
Gretchen Daily
Partha Dasgupta
Paul Ehrlich
Carl Folke
Lawrence Goulder
Geoffrey Heal
AnnMari Jansson
Bengt-Owe Jansson
Nils Kautsky
Simon Levin
Jane Lubchenco
Karl-Göran Mäler
Elinor Ostrom
Charles Perrings
Stephen Schneider
David Starrett
Brian Walker

Askö 2000. *Beijer Reprint No. 188:* Heal, G., Walker, B., Levin, S., Arrow, K., Dasgupta, P., Daily, G. C., Ehrlich, P., Mäler, K-G., Kautsky, N., Lubchenco, J., Schneider, S., & Starrett, D. (2004). Genetic diversity and interdependent crop choices in agriculture. *Resource and Energy Economics, 26*, 175–184.

The 9th Askö Meeting, 2001: Indicators

Sara Aniyar
Kenneth Arrow
Bert Bolin
Steve Carpenter
Gretchen Daily
Partha Dasgupta
Paul Ehrlich
Carl Folke
Geoffrey Heal
Michael Hoel

Nils Kautsky
Simon Levin
Jane Lubchenco
Karl-Göran Mäler
Elinor Ostrom
Thomas Rosswall
David Starrett
Brian Walker
Anastasios Xepapadeas

Askö 2001. *Beijer Discussion Paper Series No. 188:* Xepapadeas, A., Carpenter, S., Aniyar, S., Arrow, K., Daily, G., Dasgupta, P., Ehrlich, P., Folke, C., Heal, G., Hoel, M., Kautsky, N., Levin, S., Lubchenco, J., Mäler, K.-G., Ostrom, E., Rosswall, T., Starrett, D., & Walker, B. (2004). Sustainability's compass: Indicators of genuine wealth.

The 10th Askö Meeting, 2002: Dealing with Uncertainty: Policy Making in the Twenty-First Century

Sara Aniyar
Kenneth Arrow
Bert Bolin
Partha Dasgupta
Paul Ehrlich
Carl Folke
Michael Hanemann
Geoff Heal
Michael Hoel
AnnMari Jansson
Bengt-Owe Jansson
Nils Kautsky
Ann Kinzig
Simon Levin
Jane Lubchenco
Karl-Göran Mäler
Sara Aniyar
Steve Pacala
Steve Schneider
Domenico Siniscalco
David Starrett
Brian Walker

Askö 2002. *Beijer Reprint No. 180*: Kinzig, A., Starrett, D., Arrow, K., Aniyar, S., Bolin, B., Dasgupta, P., Ehrlich, P., Folke, C., Hanemann, M., Heal, G., Hoel, M., Jansson, A., Jansson, B.-O., Kautsky, N., Levin, S., Lubchenco, J., Mäler, K.-G., Pacala, S., Schneider, S., Siniscalco, D., & Walker, B. (2003). Coping with uncertainty: A call for a new science-policy forum. *Ambio, 32*, 330–335.

The 10th Askö Meeting, 2002. In this photo are the participants listed above and Anne Ehrlich, Ann Marie Heal, Christina Leijonhufvud, Carol Levin, Cristina Siniscalco, and Anna Sundbaum.

Index